BIOLOGY: BRAIN & BEHAVIOUR

Behaviour and Evolution

Springer
*Berlin
Heidelberg
New York
Barcelona
Budapest
Hong Kong
London
Milan
Paris
Santa Clara
Singapore
Tokyo*

Marion Hall
Tim Halliday (Eds)

Behaviour and Evolution

With 134 Figures

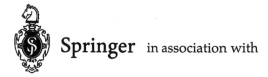

Springer in association with The Open University

Unless otherwise stated, all contributors are (or were at the time this book was written) members of The Open University

Academic Editors

Marion Hall
Tim Halliday

Authors

Marion Hall
Tim Halliday
Heather McLannahan
Frederick Toates
Terry Whatson

External Assessors

Richard Andrew, School of Biological Sciences, University of Sussex
   (Series Assessor)
Peter Slater, Zoology Department, University of St. Andrews
   (Book Assessor)

Biology: Brain & Behaviour
series

1 **Behaviour and Evolution**
2 Neurobiology
3 The Senses and
    Communication
4 Development and Flexibility
5 Control of Behaviour
6 Brain: Degeneration,
    Damage and Disorder

Library of Congress Cataloging-in-Publication Data
Behaviour and Evolution: with 134 figures/Marion Hall, Tim Halliday (eds)
Includes bibliographical references and index.
   ISBN 3-540-64752-X (pbk.)
   1. Animal behaviour. I. Hall, Marion, 1953– . II. Halliday, Tim, 1945– .
QL751.C344 1998
591.5—dc21   98-36072   CIP

Published by Springer-Verlag, written and produced by The Open University

Cover design: *design & production* GmbH, Heidelberg

Printed in Singapore by Kyodo under the supervision of MRM Graphics Ltd, UK.

ISBN 3-540-64752-X Springer-Verlag Berlin Heidelberg New York

This text forms part of the Open University *Biology: Brain & Behaviour* series. The complete list of texts which make up this series can be found above. Details of Open University courses can be obtained from the Course Reservations and Sales Office, PO Box 724, The Open University, Milton Keynes MK7 6ZS, United Kingdom: tel. (00 44) 1908 653231. Alternatively, much useful course information can be obtained from the Open University's website: http://www.open.ac.uk

2.1

SPIN 10685925   #39/3137 – 5 4 3 2 1 0

CONTENTS

PREFACE

Behaviour and Evolution, like any other textbook, is designed to be read on its own, but it is also the first in a series of six books that form part of *SD206 Biology: Brain and Behaviour,* a course for Open University students.

Each subject is introduced in a way that makes it readily accessible to readers without any previous knowledge of that area. Questions within the text, marked with a □, are designed to help readers understand and remember the topic under discussion. (Answers to in-text questions are marked with a ■.) The major learning objectives are listed at the end of each chapter, followed by questions (with answers given at the end of the book) which allow readers to assess how well they have achieved these objectives. Key terms are identified in bold type in the text; these are listed, with their definitions, in a glossary at the end of the book. Key references are given at the end of each chapter, where appropriate. A 'general further reading' list, of textbooks relevant to the whole book, is also included at the end.

The study of the brain and behaviour is an experimental science. This means that it involves the collection of observations, the formulation of specific hypotheses to explain those observations and the carrying out of experiments to test (confirm or falsify) those hypotheses. Throughout this book, these different aspects of the investigative process are emphasized, often through the use of in-text questions in which the reader is invited to engage in the process of deductive reasoning themselves. An understanding of the scientific method, as it applies to the behavioural and brain sciences, is an important aim of this book.

This book deals mainly with the behaviour of animals and some of the ways in which it is studied. This is a huge area and Behaviour and Evolution does not set out to provide a complete and comprehensive coverage of the subject. Instead, a range of topics has been selected, not only to give an informative and interesting account of animal behaviour but also to provide the background necessary to understand behaviour. Evolution is an important topic in this latter respect.

The first chapter outlines, very briefly, what behaviour is and introduces four kinds of question that behavioural scientists address to explain it: causal, developmental, functional and evolutionary. Answers to causal questions are in terms of a cause-and-effect relationship between an event that immediately precedes the behaviour and the behaviour itself. One example would be the behaviour of braking when a driver sees a red traffic light. Answers to developmental questions are also in terms of antecedent events, but in this case on a longer time scale, referring to an event earlier in the organism's life. Thus the driver brakes because they have learned earlier in life that they must stop when they see a red light. Answers to functional questions are in terms of the immediate benefits of the behaviour, specifically in the role the behaviour plays in enhancing the survival or reproductive success of the organism. Thus the driver brakes because, if they did not stop, they might have an accident, which could lead to their death. Finally, answers to evolutionary questions are in terms of the long-term history, the evolutionary antecedents of the behaviour. In the driver's case, we could say that traffic lights have proved to be a more efficient way to regulate traffic at road junctions than were previous methods such as using police. Each answer gives a different explanation of the same behaviour but all four types of answer are needed for a complete explanation.

Chapters 2 to 5 discuss in more detail each of these four aspects of the study and analysis of behaviour: Chapter 2 looks at behaviour in terms of its causation; Chapter 3 considers the contribution of inherited factors—the genes—to the development of behaviour; Chapter 4 deals with evolutionary theory and the functional aspects of behaviour; and Chapter 5 returns to developmental issues but this time with more emphasis on the environmental factors that influence behavioural development. Chapters 6 to 8 cover three very important general processes that are involved in behaviour: learning, motivation and cognition. Chapter 9 selects one particular category of behaviour, which relates to reproduction, and discusses how some of the concepts developed in earlier chapters apply to it. Finally, Chapter 10 departs from the rest of the book, which essentially deals with animals as individuals, and introduces you to the complexities of how animals interact with one another when they live in groups, looking particularly at the costs and benefits associated with group living, whether in simple aggregations or in highly-structured societies.

Before you begin to read this book, there are some important points that you should bear in mind.

1 Experiments on animals

The use of living animals in research is a highly emotive, contentious and political issue. You are no doubt aware of the strong views held by animal liberationists. There is also considerable debate among scientists concerning what kinds of experiments and procedures are acceptable and what are not. Most scientists working with animals seek to minimize any suffering that animals may experience during experiments and each researcher makes his or her own judgement as to whether the suffering caused by an experiment is justified by the scientific value of the results that the experiment yields. The ethics of animal experimentation is not simply a matter of individual judgement, however, but is a matter of concern for society as a whole. In Britain and many other countries, all researchers work within strict guidelines enforced by government; for example, the Home Office licenses all animal experimentation in the UK. Some academic societies, such as the Association for the Study of Animal Behaviour, and many institutions, such as medical schools, have Ethical Committees that oversee animal-based research. In this book, a number of experiments are described; this in itself raises ethical issues because reporting the results of an experiment may be thought to be giving tacit approval to that experiment. This is not necessarily true and it should be pointed out that some of the experiments described were carried out several years ago and a number of them would not be carried out today, such has been the shift in opinion on these issues within the biological community. Paradoxically, certain experiments carried out many years ago, such as those on the effects of maternal deprivation on young monkeys, produced such strong and distressing effects on their subjects—results that were not generally anticipated—that they have had a substantial impact on the kind of experiments that are permitted today.

2 Latin names for species

A particular individual animal belongs to various categories. If you own a pet, it may, for example, be categorized as a bitch, a spaniel, a dog, a mammal, or an animal. Each category is defined by particular features that differentiate it from other, comparable categories. The most important level of categorization in biology is at the level of the species. When a particular species of animal is referred to in this book, its Latin name is also given, e.g. earthworm (*Lumbricus terrestris*).

CHAPTER 1
INTRODUCTION

This first book in the course introduces you to the study of behaviour. The behaviour of animals, including humans, is very diverse, often very complex, and there are many aspects of it that have attracted the interest of behavioural scientists. The diversity and complexity of behaviour will become apparent to you as you read this book; in this first chapter we introduce you to the ways in which behaviour is studied. Behaviour is studied by a variety of scientists who belong to various scientific disciplines; the most important of these are ethologists, psychologists, physiologists, biochemists, anatomists and evolutionists. Each kind of specialist has a specific research agenda, a set of questions about behaviour that they are seeking to answer. The chapter begins by looking in detail at the nature of these questions.

1.1   What is behaviour?

Behaviour is, quite simply, what living animals, including humans, do. Put another way, it is what dead animals cease to do; it is a fundamental property of living things, including plants, though their behaviour falls outside the scope of this course. Behaviour typically involves movement, of the whole body, as when a predator chases its prey, or of a part of a body, as when a sleeping cat flicks its ears. Movement involves the skeleton, the muscles that move the various parts of the skeleton, and the nervous system that controls and coordinates the movements of those muscles. Movement also involves the environment, in the sense that behaviour is often directed towards something in the outside world, such as prey, a mate or a nest. Animals are able to direct their movements towards specific aspects of their external world because they have sense organs that provide them with information about the outside world that is processed in their nervous systems. Thus much of behaviour involves the interaction between an animal's 'machinery', its bones, muscles, nervous system, etc. and its outside world, such as its food, enemies and social partners.

Can a living animal ever be said to be not behaving? If behaviour is anything that involves movement, is a sleeping cat, or a heron that stands immobile waiting for an unwary fish to come by, not behaving? The answer is 'No'. Sleep is an activity in its own right; it is a response to signals from the nervous system, it involves specific actions on the part of muscles, and it occurs in response to external events, such as the onset of darkness. A heron standing immobile beside a pond is likewise involved in considerable muscular activity that ensures that it stays immobile, and it is constantly watching the water, both activities which improve its chances of catching a fish. Behaviour, then, is not an attribute of animals around which we can draw clear or obvious boundaries, other than saying that if an animal is dead it is not behaving.

It is because of the all-embracing nature of behaviour that its study lends itself, more than any other aspect of biology, to an inter-disciplinary approach. Consider the immobile hunting heron. To understand this behaviour fully we would require the expertise of anatomists and neurophysiologists to explain how a heron, unlike ourselves, can control its muscles so that it can remain utterly immobile for long periods of time. We should need the expertise of people who understand the sensory capacities and behaviour of fishes to explain why it is necessary for herons to be immobile in order to catch them.

1.1.1 Explanations of behaviour

All those scientists who study behaviour are essentially asking the question: 'Why?'. Why does an animal behave in the way it does? One of the pioneers of **ethology** (the study of the behaviour of animals in their natural environment), Niko Tinbergen, pointed out that there are four different ways of answering this apparently simple question. We can illustrate these by posing a simple question about human behaviour: Why do drivers stop when they see a red traffic light? There are four kinds of answer that we can give:

1   The red light is a stimulus to which drivers respond by applying the brake. This answer is in terms of a cause-and-effect relationship between an event that immediately precedes the behaviour and the behaviour itself; we refer to this explanation as one in terms of the **causation** of behaviour.

2   The drivers have learnt that they must stop when they see a red light. This answer is also in terms of antecedent events, but on a longer time scale, referring to an earlier stage in the individual person's life. This answer is in terms of the **development** of behaviour.

3   If drivers did not stop, they might have an accident. This answer is in terms of the immediate benefits, or **function**, of the behaviour, specifically in the role that the behaviour plays in enhancing the survival of the driver. We refer to this aspect of behaviour as its **survival value**.

4   Finally, and less obviously, we could answer that traffic lights have proved to be the most effective way to regulate traffic at road junctions, more efficient than previous methods, such as employing police. This answer is in terms of the long-term history, the evolutionary antecedents or **phylogeny** of the behaviour. The phylogeny of behaviour is an aspect of the subject that will not much concern us in this course, but it is important to realize that behaviour has an evolutionary history just as morphology (body shape and structure) does.

The important point about these four kinds of answer is that each of them, on its own, gives an explanation of behaviour, but it requires all four kinds to provide a complete explanation. Furthermore, no two of the answers are mutually exclusive; an explanation in terms of development is no better, or more complete than one in terms of causation; nor is any one kind of answer contradictory to any other.

Tinbergen, as an ethologist, was interested in the study and analysis of behaviour as such, though he was well aware of the importance of understanding underlying mechanisms at the level of nerves and muscles. To his four kinds of explanation of behaviour can be added a fifth, the *reductionist* explanation. This would analyse how the driver's behaviour is 'caused' by the actions of their leg muscles, triggered by nerves carrying signals from the brain These events could be analysed at a deeper level still, in terms of the biochemistry of nerve action and of muscular contractions.

Turning to the behaviour of animals, David McFarland has illustrated the many ways of asking questions about behaviour in a different way. A characteristic feature of birds is that their young are born as eggs which one or both parents incubate by sitting on them, thus maintaining the correct temperature for their development. We could ask the question: Why do birds sit on eggs? This can be answered in many ways, depending on where we put the emphasis.

1   Why do birds sit on *eggs*? Here, the question is about how birds recognize eggs and sit on them rather than stones or other objects. What are the specific features of eggs that elicit incubation behaviour? Do birds discriminate between eggs of their own and of other species?

2   Why do birds *sit on* eggs? Here, the question is concerned with how birds respond to their eggs in the way that they do, and not in some other way. A herring gull will incubate its own eggs but will eat its neighbour's, a few feet away, should they be left unattended. This raises questions, both about how it recognizes its own eggs—does it distinguish them by their pattern of spots or by their position on the ground?—and about its internal state or *motivation*. We assume that they sit on eggs because they're broody, and eat them because they're hungry. What, in terms of internal processes, do 'broody' and 'hungry' mean? What mechanisms enable a bird to switch quickly from one state to the other?

3   Why do *birds* sit on eggs? Here we are interested in how the behaviour of birds differs from that of other animals, like mammals, that do not sit on eggs. This is a question about phylogeny and we would seek the answer among the ancestors of birds, the reptiles, which also lay eggs. We would look for the evolutionary antecedents of incubation behaviour. Many reptiles guard their eggs but do not incubate them; some (e.g. crocodiles) lay them in places where they are incubated by the heat of the sun.

4   *Why* do birds sit on eggs? This is a question about the function of the behaviour. Birds sit on eggs in order to hatch them; if they did not the eggs would die. This is pretty obvious but there are more subtle aspects to this question. Some individuals hatch more young than others. Are they better at sitting on eggs? Do they spend more time doing it? If so, why? Are they just better parents, or is it that they are better at finding food and therefore have more time to sit on their eggs?

This example serves to show that an apparently simple behaviour pattern has many aspects and that there are many diverse and interesting questions that we can ask about it. How behavioural scientists set about answering such questions is discussed in Section 1.2.

First, we consider briefly a specific example that shows the importance of considering different kinds of explanation at the same time.

1.1.2 Bats and moths

Moths are typically active at night, when they are liable to fall prey to bats. Bats locate their prey by emitting ultrasonic sounds (sounds of very high pitch that are inaudible to human ears) and listening to the echoes of these sounds that are reflected back off objects, including insects. In effect, they use a form of sonar similar to that used in ships to detect submarines. Moths have a pair of ears, one on each side of the body, that are very sensitive to the calls of bats. Because their ears are so sensitive, and because the distance that sound has to travel from bat to moth and back to the bat is twice that from the bat to the moth, moths commonly detect a hunting bat before the bat detects them. If it does detect a bat at some distance from it, a moth takes evasive action, simply altering its flight path away from the source of danger. If, however, it does not detect a bat until the latter is very close, the moth shows a quite different pattern of evasive behaviour. It dives to the ground, often executing a series of apparently random twists and turns as it does so (Figure 1.1) (Roeder, 1963).

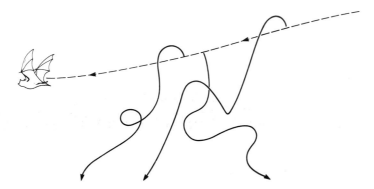

Figure 1.1   Examples of evasive flight patterns shown by moths being hunted by bats.

This dichotomy in a moth's response to hearing a bat makes sense in terms of the *function* of its behaviour. When it hears a distant bat, the bat has probably not detected it, and the optimal behaviour pattern is to fly away from the bat. If the bat is very close, however, the bat may have detected it and simply flying away may not prevent its demise. By diving in an erratic fashion it may escape the bat, which sends out sound in a fairly narrow beam in front of its head and so can only detect objects directly in front of it.

What is going on in terms of *causation*? The moth's ears are very simple, each consisting essentially of two sound detectors that have slightly different thresholds. What that means is that it requires a slightly louder sound to activate one detector rather than the other; put another way, one detector is more sensitive than the other. The sound of a distant bat will be sufficient only to activate the lower-threshold detector; that of a bat close by will activate both detectors. Thus, the moth's behaviour varies according to how many detectors are activated at a given time.

The particular properties of the sound detectors in the moth's ear are interesting in their own right, but their significance does not become apparent until we consider what role the moth's ear plays in its life. When both the ecological situation in which moths live and the exact nature of the threat to them posed by hunting bats are understood, the properties of their ears make much more sense. This illustrates the importance of considering more than one question about behaviour, in this case causation and function, at the same time, and of maintaining an interplay between different kinds of explanation. It also illustrates the importance of an interdisciplinary approach, in this instance between neurophysiology and ethology. We shall return to the ears of moths later in the course.

1.2   Approaches to the study of behaviour

There are two fundamental components to the study of behaviour: observation and experiment. To begin to understand behaviour, it is necessary to describe it. This involves making careful observations so that a detailed description of everything that an animal does in a given context can be built up. It is particularly important that the full range of variation in a particular pattern of behaviour is described, as this is often very illuminating. For example, the fact that moths show different evasive behaviour when bats are close, compared with when they are distant, raises particular questions about how the moth's ear actually works.

By observing patterns of behaviour repeatedly, it is possible to build up a picture of how often particular actions occur and what the relationship between different actions is. It requires many observations of moths flying about at night to establish that there is a relationship between the nature of their evasive behaviour and the proximity of bats. In building up this picture, we start to have ideas about the association between one event and another. These associations may be in terms of the causation, the development or the function of the behaviour. The next step in the analysis of behaviour is to formulate these ideas as specific hypotheses, and then to devise experiments to test those hypotheses. The following examples illustrate how different questions about behaviour can be studied experimentally.

1.2.1   Causation: courtship in newts

Smooth newts (*Triturus vulgaris*) are common inhabitants of ponds, to which they return every spring to breed. Males show elaborate **courtship** behaviour, in which they perform various movements, called **displays**, in front of females. (In ethology, the term *courtship* refers to male and female behaviour patterns that precede and accompany the act of mating.) If a male succeeds in attracting a female, she approaches him. He then turns and creeps away from her, waving his tail from side to side, and she touches his tail with her snout (Figure 1.2, *overleaf*). Repeated observations of newt courtship reveal that this tail-touch by the female is almost always followed, on the part of the male, by the release of a packet of sperm, contained in a membrane, called a spermatophore. This leads to the hypothesis that the female's tail-touch *causes* the male to deposit a spermatophore. This hypothesis has been tested by the following experiment, carried out by Tim Halliday (1975).

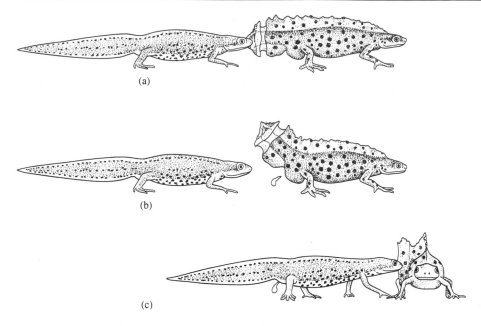

(a)

(b)

(c)

Figure 1.2 Spermatophore transfer in the smooth newt (*Triturus vulgaris*). (a) The female touches the male's tail; (b) he deposits a spermatophore; (c) the spermatophore attaches to the female's genital opening.

A female is placed in a special harness called a 'strait-jacket', such that she can be moved about by the experimenter. She is moved towards a male, which begins to display to her. After a while she is made to approach the male, which turns, creeps and waves his tail. At this point the experimenter either moves the female so that her snout touches the male's tail or holds her back so that he receives no tail-touch.

☐   If the hypothesis is correct, what will the male do in each of these situations?

■   He should produce a spermatophore when his tail is touched, and not do so when it is not.

The results of many experiments are presented in Table 1.1

Table 1.1 Results of a series of experiments in which the causal relationship between the tail-touch by the female smooth newt (*Triturus vulgaris*) and spermatophore deposition by the male was investigated.

	Number of times spermatophores deposited	Number of times spermatophores not deposited
Tail-touch given	31	1
Tail-touch withheld	4	31

☐   Do the results in Table 1.1 support the hypothesis or not?

■   They do. In the majority of times in which a tail-touch was given, the male deposited a spermatophore; in the majority where it was withheld, he did not.

We can conclude, therefore, that tail-touch is the stimulus that elicits spermatophore deposition. Note, however, that there were some trials when this causal explanation does not seem to hold. Some males deposited a spermatophore without receiving a tail-touch; one failed to do so even when its tail had been touched. Do we dismiss these as aberrations or can we find some way of explaining their occurrence?

By looking back at the detailed records of the behaviour of males in these 'aberrant' trials, it was possible to establish that, in those trials in which males deposited a spermatophore without having received a tail-touch, the male showed evidence of being especially sexually active (his display rate was very high). In contrast, the one male that failed to deposit despite receiving a tail-touch was less sexually active.

☐   What do these observations suggest to you about the causation of the male's behaviour?

■   The male's behaviour is influenced by his readiness to court females, i.e. his motivation to mate.

Males with high sexual motivation will deposit sperm without receiving a stimulus from the female; those with low sexual motivation sometimes fail to respond to the stimulus. In fact, such males require two, three or even four tail-touches before they will deposit. This interaction between external stimuli and an animal's internal state is taken up in Chapter 7 of this book.

1.2.2   Function: bird song

Males of many species of bird spend a large proportion of their time singing elaborate songs, especially during the breeding season. It has been suggested that song serves two primary functions, attraction of females and exclusion of rival males from a male's **territory** (defined as a defended area). Support for the first of these hypotheses comes from two kinds of evidence. First, in several species, such as the sedge warbler (*Acrocephalus schoenobaenus*), the amount of time that a male spends singing falls dramatically as soon as he has attracted a mate. Second, in the great tit (*Parus major*), a male shows a resurgence in singing if his female is caught and removed.

The second hypothesis, that song acts as a 'keep out' signal to other males, has been tested experimentally by John Krebs (1977). He used a technique called a 'playback' experiment, in which recordings of male songs are broadcast, in the birds' natural habitat, through loudspeakers. Working in a small wood outside Oxford, Krebs and his colleagues first removed all eight males that had set up territories in the wood in the early spring (Figure 1.3a, *overleaf*). They then divided the wood up into three areas. In one area, no recorded sounds were played, in another recordings of great tit song were broadcast through loudspeakers and, in the third area, speakers broadcast a 'control sound', consisting of a tune played on a tin whistle (Figure 1.3b). They then observed the wood continuously for several days and recorded when male great tits moved in and set up territories. Figure 1.3c shows the pattern of reoccupation of the wood by new males.

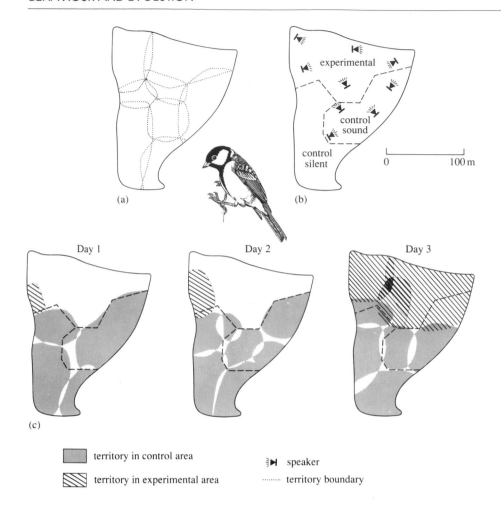

territory in control area   ⊯ speaker

territory in experimental area   ⋯⋯ territory boundary

Figure 1.3 Song as a 'keep out' signal in the great tit (*Parus major*). (a) The pattern of eight territories existing in the wood before territorial males were removed. (b) The wood divided up into three areas, two of which contained loudspeakers through which recorded sounds were broadcast. (c) The pattern of reoccupation of the wood by new male great tits over the next three days.

☐ Do the results shown in Figure 1.3c support the hypothesis that great tit song serves the function of a 'keep out' signal?

■ Yes, they do. That part of the wood in which great tit songs were broadcast was reoccupied more slowly than those in which no sound or a control sound were played.

☐ What do you suppose was the purpose of using the sound of a tin whistle as a control sound?

■ To eliminate the possibility that great tits might be deterred from occupying a part of the wood in which sound of any kind was being played, or even by the mere presence of tape recorders and loudspeakers.

1.2.3   Development: bird song

Birds are well known for the variety and complexity of their songs. A feature of bird song is that it is highly **species-specific**; each species has a characteristic song pattern, as a result of which an experienced ornithologist can identify a bird by its song alone. How does bird song develop? Is each individual hatched with the ability to sing the song characteristic of its own species or does it have to learn that song during its early life? This takes us into an important, and often controversial area of behavioural research, the question of whether a particular aspect of behaviour is inherited, or whether it is learned. This is often referred to as the *nature–nurture debate* ('nature' referring to inheritance, 'nurture' to learning), to which we will return a number of times in the course. Konrad Lorenz (with Niko Tinbergen, one of the founders of ethology) suggested that one could separate the two factors by means of a **deprivation experiment**. In this, the developing animal is kept under conditions such that it does not experience the environmental stimuli that the experimenter assumes are necessary for learning a particular pattern of behaviour. If it nonetheless does develop that behaviour pattern, it must be inherited; if it does not, it must be learned.

In the case of bird song, many studies have tested the hypothesis that male birds learn to sing by hearing the song characteristic of their species early in life. This can be tested by rearing young birds in acoustic isolation from adult males of their own species. The results of such experiments vary from species to species. For most species early experience of hearing song is necessary. In some species, such as zebra finches (*Taeniopygia guttata*) and Bengalese finches (*Lonchura striata*), males that are reared in the presence of adult males of other, but not of their own species, will learn the song of their 'foster' species. If, however, they are reared with both, they develop their own species' song; this suggests that, though learning is vital, there is some innate predisposition to sing their own species' song rather than another's. Many of these experiments have also shown that birds are not equally likely to learn songs at all ages; for many species, there is a **sensitive period**, a specific period of time during which they are particularly receptive to song and during which song learning occurs.

The rationale behind deprivation experiments may seem straightforward, but the nature–nurture issue is not so simply resolved. A major problem is that, in setting up a deprivation experiment, a researcher has to pre-judge what is the particular experience that is relevant to the behaviour pattern in question. For example, some birds can sing their own species' song, even if they have never heard it before. Others require exposure to their species' song if they are to sing it correctly. Thus, the development of bird song may require at least two kinds of experience: hearing other birds sing and hearing oneself sing. All that a deprivation experiment can tell us is that the specific experience that has been manipulated does or does not influence development; it cannot establish that a behaviour pattern is not influenced by external factors at all during development.

What deprivation experiments do, in effect, is to reveal the properties of a system in the absence of a particular stimulus. They are analogous to experiments on the brain in which parts are removed (a procedure called lesioning) which show how the brain works when a part of it is missing.

The term 'innate'

The word '**innate**' came into the ethological vocabulary largely because the words 'instinct' and 'instinctive' became too imprecise for scientific usage. This is due to the fact that they are used in four different ways:

1   To mean 'inherited' or 'present in the genes'. Such behaviour is therefore characteristic of the species and stable from generation to generation.

2   In connection with motivation, as in 'the parental instinct motivates a bird to build a nest'.

3   To label fast and appropriate responses that have been acquired as a result of learning, as in 'tennis players know instinctively which shot to play'.

4   To label a behaviour pattern that is performed correctly the first time it is performed, apparently without any experience or practice, e.g. sucking at the breast by human babies.

The term 'innate' has, unfortunately, also been used in more than one sense, principally senses 1 and 4. In this course, we define it in the sense in which it is used in 4. A better term for this might be 'congenital' but this is not widely used in ethology. Note that sense 4 is often taken to imply sense 1, i.e. if a behaviour pattern is present at birth it must be inherited, but this is not necessarily so, a point to be discussed fully in Chapter 5.

1.2.4   Ethology and psychology

The behaviour of animals has been studied, in rather different ways, primarily by two groups of behavioural scientists, ethologists and psychologists. Ethologists are primarily concerned with the behaviour of animals in the context of their natural environment. Consequently, they typically work 'in the field' (meaning the natural environment rather than the laboratory) observing behaviour in such a way that the behaviour of their subjects is as little affected by their activities as possible. Ethologists frequently also work in the laboratory but, when they do so, they generally create an environment in which the particular behaviour patterns with which they are concerned are performed as normally as possible. Much work on fishes, for example, is carried out in aquaria where fish can behave much as they do in nature. Psychologists adopt an essentially experimental, laboratory approach, in which the external world of the animal is carefully and systematically controlled and in which the behaviour of the animal can be observed in great detail, and many aspects of its physiology can be monitored precisely. The difference between the two approaches has been parodied thus: psychologists put the animal in a box and watch it, ethologists put themselves in a box and look out at the animal.

Because of their emphasis on behaviour in a natural environment, a major concern of ethologists is the function of behaviour. Only in a natural setting, in which animals have evolved, can we see what role a particular pattern of behaviour plays in an animal's life. Ethologists are also concerned with development and causation, but it is much more difficult for them to study these in the field, because the natural environment is so complex and variable. The psychologist, working with animals in an artificial environment, can control to a greater extent exactly what the animal experiences during its life. As a result, psychologists have built up a much more thorough and detailed picture of the mechanisms involved in learning and motivation than field-based ethologists could ever hope to do.

As field biologists, ethologists are very aware of the diversity of animals. As a result, they work on a very wide variety of animals, and have built up a substantial literature of studies on a vast array of species. Their concern is to discover the ways in which each species relates to, and has evolved in relation to its particular environment. Because animal species are so varied and diverse in their habits and in the environments in which they live, it is difficult to establish general principles that apply to the behaviour of all species. Psychologists, in contrast, are more concerned with processes, such as those involved in learning and motivation, and have concentrated their efforts on relatively few laboratory species, such as rats. In the past, there was a tendency for some psychologists to assume that mechanisms of learning revealed in rats are common to all animals, an assumption that an ethologist, aware of the diversity of animals, would be very wary about.

The approaches of psychologists and ethologists are essentially complementary, and information gained by one approach is essential to the other. This creative interplay between different disciplines is the essential rationale for this, interdisciplinary, course. There is, however, a certain tension between the two disciplines, which has surfaced from time to time in heated debates and controversy. Some ethologists have dismissed the findings of psychology because they are based on behaviour that is observed in unnatural settings. Psychologists have accused ethologists of ignoring the importance of learning in behaviour and of placing too much emphasis on the concept of instinct. These disagreements have been very marked at times but, in recent years, there has been an increasing trend for the two disciplines to respect one another's approach to behaviour, to integrate their findings, and to appreciate that there is much that each has to learn from the other.

Summary of Chapter 1

The behaviour of an animal consists primarily of movements made by parts or all of its body, though there are situations in which immobility is an important aspect of behaviour. In the scientific study of behaviour, various questions are posed; these relate to the causation, the development, the function and the evolution of behaviour. Answers to these different types of question are not mutually exclusive but complement one another; a complete understanding of behaviour requires answers to all four kinds of question. Behaviour can be very complex and variable and can be seen to be coordinated in response to features in an animal's environment. The moth, for example, has a pair of simple ears that act as bat detectors and enable it to take evasive action that is appropriate to its distance from a bat. Behaviour is studied primarily by observing and describing it. This leads to experiments, in which specific hypotheses about its causation, development and function are tested. Research by ethologists and psychologists into the behaviour of animals has involved different approaches that complement one another.

Objectives for Chapter 1

When you have completed this chapter, you should be able to:

1.1 Define and use, or recognize definitions and applications of each of the terms printed in **bold** in the text.

1.2 Appreciate that the analysis of behaviour involves multiple approaches, and to recognize examples of these approaches.

1.3 Appreciate the importance of an experimental, hypothesis-testing approach to the analysis of behaviour.

1.4 Distinguish between causal, functional, developmental and evolutionary explanations of behaviour. (*Question 1.1*)

Question for Chapter 1

Question 1.1 (*Objective 1.4*)
The following are four statements concerning bird song:

(i) Male song attracts females into a male's territory and deters intrusion by other males.

(ii) Male birds sing mainly in the spring, when there is a high level of sex hormones in their blood.

(iii) An individual male includes a number of specific features in his song that were characteristic of the song of his father.

(iv) If his mate dies or is removed from his territory, a male shows a marked increase in his song output.

Which of the statements (i) to (iv) refers to (a) the causation, (b) the development, (c) the function of bird song?

References

Halliday, T. R. (1975) An observational and experimental study of sexual behaviour in the smooth newt, *Triturus vulgaris* (Amphibia: Salamandridae). *Animal Behaviour*, **23**, pp. 291–322.

Krebs, J. R. (1977) The significance of song repertoires: the Beau Geste hypothesis. *Animal Behaviour*, **25**, pp. 475–478.

Roeder, K. D. (1963) *Nerve Cells and Insect Behaviour*, Harvard University Press.

CHAPTER 2
ACTIONS AND REACTIONS

2.1 Introduction

This chapter is concerned with the analysis of behaviour in terms of its *causation*. What this means was outlined briefly in Chapter 1 but, before going any further, it is necessary to recall exactly what kind of causality is being considered. In a general sense, to say that a particular pattern of behaviour is *caused* by something implies that that something preceded the behaviour in time. In considering the behaviour of animals, ethologists are concerned with processes that occur on more than one time-scale. An animal's current behaviour is influenced by events that are occurring over seconds and minutes, by events that occurred days, months or years earlier in its life, and by its evolutionary history. Analysis of these latter two aspects was categorized in Chapter 1 as being concerned, respectively, with development and evolution. This chapter is essentially concerned with the 'here and now' analysis of behaviour, what Aristotle called its *efficient* cause, what is now called its *immediate* or *proximate* cause.

To say that the behaviour of animals is caused by something, means that a particular pattern of behaviour, called a **response**, is associated closely in time with some specific object or event in the animal's environment, called a **stimulus** (plural: stimuli). A frog will turn towards and then flick its tongue towards a fly that comes close to it; the fly is a stimulus that *elicits* a feeding response. Stimuli may be objects, like food items, enemies, potential mates or social partners, or they may be events, like the onset of rain, the time of day or the season of the year. They may, like these examples, be external to the animal, or they may be internal, and relate to physiological changes which, for the moment, we can loosely label by such terms as hunger, thirst, aggression or fear.

2.2 Reflexes: animals as automata

The behaviour of jumping spiders in response to external stimuli has been characterized, somewhat facetiously, by someone who has worked on them for many years as follows: if it is small, eat it; if it is large, run away from it; if it is intermediate in size mate with it. The behaviour of jumping spiders is not this simple; nor is most animal behaviour. As this and later chapters will show, animals are not automata that respond in an entirely predictable fashion to external stimuli. There are, however, examples of behaviour that are this simple, where a specific stimulus always elicits a particular response; such a system is called a **reflex**. A classic example is the knee-jerk response, illustrated in Figure 2.1.

When a doctor taps your knee, it straightens. In healthy individuals, this is an invariable and entirely predictable association of external stimulus and behavioural response. This description of the behaviour is, for some purposes, a

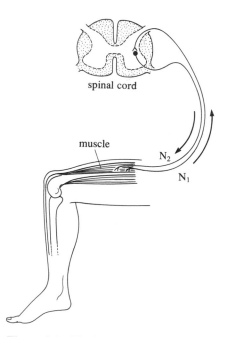

Figure 2.1 The knee-jerk reflex. A tap just below the knee generates a signal in the form of nerve impulses via two nerve cells (N_1 and N_2), the first to the spinal cord and the second to muscles in the leg which contract.

sufficient level of analysis. A biologist could go on and consider, for example, the function of the response, how it enhances an individual's survival, without needing to know anything else about it. This is often referred to as the 'black box' approach to behaviour, which seeks to explain behaviour in terms of directly observable phenomena without seeking to understand underlying mechanisms. A researcher may, however, be interested in the underlying mechanisms involved, how the response works at the level of nerves and muscles. As shown in Figure 2.1, the knee-jerk reflex involves a loop that begins with cells that detect the knee-tap and which send nervous signals, via the spinal cord, to muscles in the leg that contract to straighten the knee. An interesting feature of the knee-jerk reflex is that the nervous pathway that controls it does not involve the brain. This kind of analysis provides a reductionist explanation, in that it explains events at one level (behaviour) in terms of events at a lower level of organization (nerves and muscles).

Black box and reductionist approaches both have their strengths and weaknesses, but for the moment it is sufficient to think of them as a reflection of the fact that different researchers are interested in different things. The German biologist J-P. Ewert, who has studied the responses of toads to visual stimuli, wrote:

> In response to specific stimuli one can repeatedly elicit predictable reactions, such as snapping at prey, fleeing from an enemy, clasping during courtship and making particular wiping motions after tactile stimulation. (The fickle European frog, in contrast, undergoes short-term changes in motivation and is not suitable for behavioural experiments.) (Ewert, J-P. (1974)

Where Ewert's interests end, those of other biologists begin; researchers interested in the physiological basis of hunger, for example, would prefer frogs to toads as experimental subjects, because they do show changes in motivation (this will be discussed in Chapter 7).

Although the concept of a reflex provides an adequate explanation for only a small part of the behaviour of animals, it is an important concept for two reasons. First, reflexes have provided valuable material for the analysis of the neurobiological basis of behaviour, as in the example of the knee-jerk reflex. For those biologists, like Ewert, who are interested in the interaction between sense organs, the nervous system and behaviour, they are often deliberately chosen because they are so reliable. Secondly, the reflex concept can be used as a model against which to compare variation in behaviour, and so point the way towards other mechanisms that are involved. Chapter 1, for example, described how variation in a male newt's response to having his tail touched by a female can be related to a male's sexual motivation.

2.3   The senses and behaviour

If animals are to respond to objects and events in their external environment, they must be able to perceive them. Information about the environment, or sensory input, is obtained through a variety of sense organs, such as the eye, the ear, the nose and touch **receptors** (a sensory cell that responds to a particular kind of stimulus). The nature of these is discussed more fully below. A major reason why the reflex concept is an inadequate explanation for much of the behaviour of animals is that animals are typically not simply passive receivers of external stimuli, but interact with their environment in complex ways. There are three major aspects to this interaction.

First, sense organs typically gather much more information than the animal is able to use. The human eye, for example, contains about 120 million light receptors, each capable of discriminating between at least ten levels of light intensity. It is inconceivable that, for every possible combination of sensory inputs (10×120 million), there is a reflex behavioural response. What actually happens is that the information gathered by sense organs like the eye is filtered, in the eye itself and in the nervous system, so that the animal responds selectively to those aspects of its sensory input that are of most relevance to it. There are two general, but not mutually exclusive, ways in which this filtering is achieved. The structure or neurophysiology of a sense organ may be such that it responds more strongly to certain kinds of sensory input than others. In general, animals with well-developed eyes respond immediately to large, looming objects that form a rapidly expanding image on the retina. Such objects are likely to be dangerous, like an attacking predator or an approaching bus, and elicit rapid evasive behaviour. More specifically, the eyes of frogs and toads are particularly good at detecting small, moving objects that are potential prey. Also, animals vary their response to sensory input according to what they are currently doing and their current internal state. Thus, many animals are particularly responsive to food objects when they are hungry. In a classic experiment, David Lack showed that a male robin (*Erithacus rubecula*) will vigorously attack a stuffed robin and that it shows a similar response to a bunch of red feathers attached to a twig (Figure 2.2). Red feathers do not, however, elicit attack in a male if they are growing on the breast of his mate.

Second, the sensory input that an animal receives changes continuously, not only because the external world changes, but also because the animal's relationship with that world changes as a consequence of the animal's own behaviour. Any movement by the animal, even a slight movement of just its eyes, will radically change the sensory input to its eyes, for example. An animal receives two kinds of sensory input, that which derives from events in its external environment, and that which results from its own actions. If it is to be able to respond appropriately to external stimuli, it must be able to separate these two processes. A distinction must thus be made between a simple reflex, in which the animal is essentially passive with respect to its environment, and a more interactive relationship between the animal and its environment.

The information that an animal receives is of two kinds: **exafferent** information that comes from changes in the environment and **reafferent** information that is due to changes that result from the animal's own activities. The **reafference principle** states that an animal can only respond to objects and events in its environment if its nervous system is able to distinguish between exafferent and reafferent information. If someone is moving around a room, they perceive it as stationary because their nervous system allows for the reafferent effect of the movements of their body and eyes on the changes in sensory input to their eyes that result from those movements. If an object, such as a cat, moves in the room, they detect this because it also provides a change in exafferent input. (When someone is drunk their ability to separate out exafferent information lets them down and the room itself appears to move.)

(a)

(b)

Figure 2.2   (a) A male robin attacking a stuffed robin. (b) This response can also be elicited by a bunch of red feathers.

Third, the perception of information is commonly not a passive process, but an active one. Animals investigate and explore their environment; they examine or sniff objects. In some instances, they are themselves the ultimate source of the incoming information, as in the example of echolocating bats that emit sounds and listen to the echoes. Thus animals commonly actively seek stimuli and are not simply recipients of them. The kind of stimuli that they seek at any one time is likely to be related to their immediate needs, as when a food-deprived animal goes in search of food.

2.3.1 The sensory world

It is natural for humans to assume that what they perceive with their eyes, ears and other senses is a complete representation of the external world. It is not, but is only that part of the world that they perceive and it is very different from that which is perceived by other species. It is always very important when studying the behaviour of other animals to bear in mind that they may live in a very different sensory world to that of humans. Birds can, in general, see and hear much better than humans can, but have a less well-developed sense of smell. Many birds are able to perceive the earth's magnetic field, which humans cannot, and many mammals such as dogs live in a sensory world that is dominated by their highly-developed sense of smell. Insects typically are able to see ultraviolet light to which humans are insensitive and, as a result, probably perceive flowers and other insects in a very different way from the way in which people perceive them (Figure 2.3).

Figure 2.3 Flowers and insects as seen under normal light (above) and through an ultraviolet detector (below).

2.3.2 Categories of senses

The term **sense**, often also referred to as a 'sensory modality', refers to the reception of a particular kind of energy or physical or chemical stimulus. The various kinds of senses are divided into three categories:

1 Enteroceptive senses. Sense organs within the interior of the body, called **enteroceptors**, monitor various aspects of the animal's internal state such as temperature, blood sugar, oxygen, etc.

2   Proprioceptive senses. Sense organs throughout the body, particularly in the limbs, monitor the position of one part of the body relative to others. These **proprioceptors** provide the brain with continuous information about the position and movement of all parts of the body.

3   Exteroceptive senses. The **exteroceptors** are those sense organs that monitor the external environment; in humans they provide the five senses, vision, hearing, touch, taste and smell. Other species may lack some of these senses; others have other senses, such as electromagnetic sensitivity.

2.4   The organization of behaviour and its analysis

Before any analysis of the extent to which behaviour consists of responses to specific stimuli can be undertaken, it is necessary to find a way to describe it and break it up into units of analysis. This is no easy task, first because behaviour is so complex and variable and, secondly, because there are no immediately obvious units. As discussed in Chapter 1, behaviour is essentially concerned with movement, and so one possibility would be to break patterns of behaviour up into individual limb movements, or into individual muscle contractions. It is immediately apparent, however, that such an approach would lead to descriptions of behaviour that would be ludicrously complex and cumbersome. Consider, for example, a relatively simple pattern of behaviour, the feeding of a pond snail. Snails move slowly across the algal-covered wall of an aquarium, rasping at the algae with a file-like 'tongue' called a radula, moving their head from side to side as they move along. Detailed analysis of this behaviour pattern has revealed that it involves 25 different pairs of muscles contracting in particular combinations and sequences. In practice, researchers adopt a more pragmatic approach, dividing behaviour patterns up into units that are suitable for the kind of analysis that they wish to do. Thus, someone interested in the physiological basis of snail feeding would have to study individual muscle contractions, but someone interested in whether snails browse on algae with maximum efficiency would look at individual scraping movements of the radula, turns of the head, and the speed of motion of the body.

2.4.1   Fixed action patterns

An early attempt to address the problem was made by Konrad Lorenz, who proposed the concept of a **fixed action pattern**. Lorenz drew attention to the fact that animals belonging to different species typically perform behaviour patterns that are very stereotyped and that are highly specific to their species. Thus, among birds, individuals sing species-specific songs, perform species-specific displays and build nests of a characteristic structure. Lorenz argued that such behaviour patterns are just as diagnostic of a particular species as its morphology and its plumage. Lorenz argued that fixed action patterns (FAPs) are pre-programmed sequences of behaviour that, once triggered by some stimulus, called a releaser, are completed without any further input from the environment. Lorenz's concept was more than a device for describing behaviour in terms of biologically meaningful units. It also included the idea that FAPs are controlled by fixed 'programs' encoded in the nervous system, and, most controversially, that these programs were inherited

The FAP concept has now fallen out of favour, following criticism of all of its aspects. The general nature of these criticisms is as follows:

1   The 'fixety' of behaviour is more apparent than real. Behaviour patterns that appear highly stereotyped often turn out, when analysed in detail (for example, by means of video recordings), to be quite variable. One ethologist has proposed that the FAP concept be replaced by the MAP (**modal action pattern**), which describes behaviour patterns in terms of the extent to which they vary about an average, or modal value.

2   Some apparently stereotyped patterns of behaviour have proved, on detailed analysis, to be influenced by events in the external environment occurring during their performance.

3   Detailed analysis has not always supported the idea that the nervous system contains encoded sequences of activity that control behaviour. We will discuss this further later in this chapter.

4   The distinction between learned and inherited, made so strongly by Lorenz, is invalid. In particular, there is no reason to suppose that because a behaviour is stereotyped, it must be inherited. For example, many animals, when kept in confinement in zoos or farms, develop repetitive, stereotyped behaviour patterns, called stereotypies. These include jumping up and down, chewing the bars of a cage, and pacing up and down along a fixed path. A feature of stereotypies is that they are highly specific to individuals.

☐   Does this support the hypothesis that stereotyped behaviour is inherited?

■   No, it does not. If different individuals develop different stereotypies, such behaviour patterns must be developed by each individual during the course of its life.

Whatever the limitations of Lorenz's ideas, his concept of the FAP has been very important in the development of the study of behaviour. For example, his suggestion that behaviour is controlled by fixed 'programs' in the nervous system caused neurophysiologists to look for them; even if such programs prove illusory in many species, much is learned about the neurophysiological basis of behaviour as a result of such research. In fact, as described below, some behaviour patterns are controlled by the nervous system much in the way that Lorenz suggested.

One example of an FAP that was quoted by Lorenz is egg-retrieval in birds such as the greylag goose (*Anser anser*) (Figure 2.4). If an egg rolls out of a goose's nest, or if one is placed just outside by an experimenter, the goose will reach out its neck, hook its bill over the egg and draw its head back into its body until its bill is between its legs, so that the egg is pulled into its nest. These movements are highly stereotyped, such that, should the egg roll away to one side, the goose still completes the movement, pulling its bill into its body before reaching out to start again. During egg-rolling, however, the goose counters any sideways movement by the rolling egg by making small sideways movements with its bill.

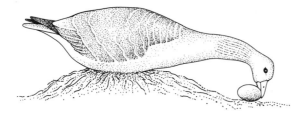

Figure 2.4   A greylag goose retrieving an egg.

☐   On the basis of this description, to what extent do you think it appropriate to call this behaviour a fixed action pattern?

■   The major element of the behaviour, reaching out the bill and pulling it back, appears to be highly stereotyped and, once begun, is always completed. The fact that the goose adjusts the position of its beak to counter sideways movements of the egg, however, shows that the behaviour is not completely stereotyped, but is varied according to changes in the external stimulus.

Lorenz argued that egg-rolling behaviour incorporated an FAP, the longitudinal component of the movement, with a variable orientating component, the lateral movements in response to sideways movements of the egg.

Whether or not a pattern of behaviour like egg-rolling, or a component of it, is an FAP as defined by Lorenz, depends ultimately on the way in which it is controlled by the nervous system. Lorenz's concept incorporated the idea that an FAP is controlled by a fixed series of nervous events that occur in a stereotyped fashion once triggered by a stimulus. We know nothing about the neurophysiology of greylag goose behaviour and so we cannot resolve whether egg-rolling is an FAP in this very strict sense or not. There are animals, however, with nervous systems that can be readily studied and, in some instances, something akin to Lorenz's concept has been demonstrated. One example is the escape response of the sea slug *Tritonia* (Figure 2.5, *overleaf*).

Sea slugs are aquatic molluscs and, like their relatives, slugs and snails, they generally crawl over hard surfaces on their large foot. They are preyed upon by starfish and, if a sea slug contacts a starfish, it shows a stereotyped escape response, lifting off the substrate and swimming away by means of a series of alternating up and down flexions of its entire body lasting about 30 seconds. This response could be a true FAP, controlled by a fixed sequence of nerve impulses from the sea slug's brain. Alternatively, it may be that each downward flexion of the body is a response to the preceding upward flexion, and *vice versa*, and that the behaviour ceases when the sea slug perceives that it is clear of danger. To differentiate between these two possibilities, neurophysiologists have dissected out the brain of *Tritonia*, cutting all the nerves that connect it to the body, and placing it in a dish where its nervous activity can be recorded. They found that, if the brain is stimulated by a slight electric shock, it produces a sequence of nerve impulses from its various nerve stumps that is exactly the same as that produced by the brain of an intact sea slug during the escape response. Such observations support the hypothesis that the escape response is controlled by a fixed 'program' of nerve impulses and refute the alternative hypothesis that each muscular contraction is a response to the preceding one.

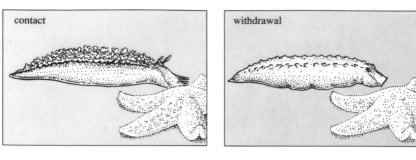
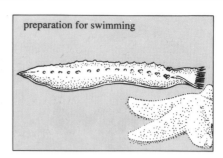
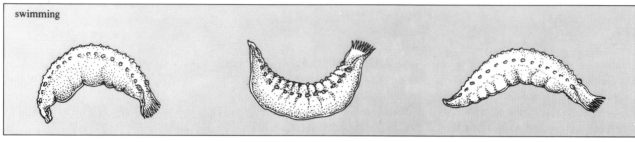

(a)

Figure 2.5 The escape response of the sea slug *Tritonia*. (a) The escape behaviour. Swimming involves muscles contracting alternately along the animal's back and underside. (b) The 'program' of nervous activity that accompanies the behaviour for a nerve cell causing contractions along the back (bottom graph) and underside (top graph). Each thin vertical line represents a single nerve impulse; the black areas represent periods when impulses are so frequent that they cannot be distinguished on the graph. (c) The brain of *Tritonia* dissected out.

There are a number of examples of quite complex behaviour sequences in which it has been demonstrated that fixed sequences of brain activity control behaviour to a large degree; most of these examples come from invertebrates such as molluscs and insects. It is rare, however, that external stimuli play no part in controlling the behaviour. For example, the contractions of the 25 pairs of muscles that control feeding in pond snails are controlled largely by a 'program' in the nervous system but, should a snail encounter a particularly obdurate food item, its grazing movements are modified such that it can 'scrape harder' at it. The elaborate series of movements by which a moth emerges from its pupa can last several hours and has been shown to be controlled by a fixed pattern of brain activity.

Although there is some evidence to support certain aspects of the FAP concept, it does not help much in the analysis of behaviour at the level of the whole animal in its environment. Observation of a stereotyped pattern of behaviour may raise questions about its immediate causation, or its function, and a researcher would not wish to be distracted by having to conduct the many years of research that it

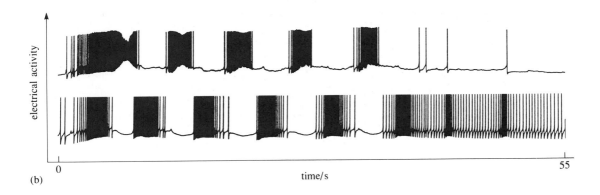

(b)

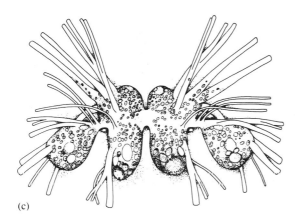

(c)

would take to establish whether or not it was controlled by a stereotyped series of nerve impulses, or whether its development in the animal's life is or is not influenced by learning. In practice, the ethological analysis of behaviour consists of making repeated observations of behaviour, attempting to identify behavioural elements that can be regarded as more or less stereotyped 'units', and trying to relate the occurrence of such units to external events.

2.4.2   Sequences of behaviour

Once a number of discrete and defineable behavioural acts have been identified, they can be analysed to see how they are ordered in time, to produce a sequence of behaviour. Like the behavioural acts themselves, behavioural sequences vary one to another in the degree to which they are stereotyped. Because ethologists are interested in the extent to which patterns of behaviour are performed in response to external stimuli, they have also to record carefully any external events that are occurring at the same time as the behaviour under consideration. These commonly relate to the behaviour of another animal, either of another species, as when a predator pursues its prey, or of the same species, as when animals are fighting or are engaged in courtship and mating. The resulting description of behaviour is sometimes referred to as an **ethogram**. Figure 2.6 illustrates a short sequence of behaviour that was discussed briefly in Chapter 1: it is part of a larger sequence of behaviour, the courtship of smooth newts.

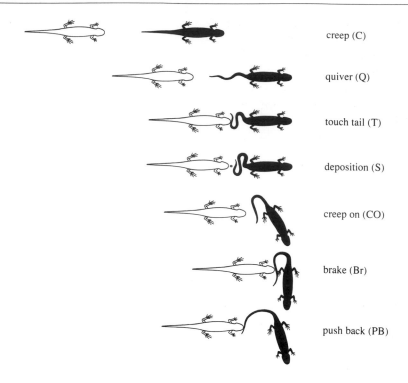

Figure 2.6 The spermatophore transfer phase of courtship in the smooth newt (*Triturus vulgaris*). The typical sequence proceeds down the page; the male is shown in black.

Consider first the sequence of male actions Creep–Quiver–Spermatophore deposition (C–Q–S). Repeated observations of pairs of courting newts reveal two things. First, this sequence is very stereotyped; it may occasionally be C–Q–C–Q–S, but it is never C–S, or S–Q, for example. Secondly, there is a strong association in time, or temporal correlation, between Q, the female's tail-touch (T) and S, such that the most common sequence of events is Q–T–S. This leads to the hypothesis that T is the stimulus that elicits the response S.

☐ Can you think of an alternative hypothesis?

■ That S always follows Q and that T just happens normally to be interposed by the female.

As described in Chapter 1, these two hypotheses can be tested and differentiated between by an experiment in which the female's behaviour is controlled by the experimenter, such that T is given or witheld systematically.

☐ Which hypothesis was correct? (See Table 1.1 in Chapter 1.)

■ That S is normally performed in response to T, though there is some variation, such that Q–S is sometimes observed, and that occasionally S does not follow T.

Considering the rest of the sequence shown in Figure 2.6, the same kind of procedure can be carried out on other parts of the sequence, to produce the results shown in Figure 2.7. This shows that the male's sequence of behaviour incorporates two kinds of causation. Some transitions from one act to another, such as C–Q and S–CO occur without input from the female, whereas both S and PB are responses to the female act T. Those parts of the male sequence, such as S–CO–Br, that are independent of input from the female are akin to FAPs, though we know nothing of their neural basis or their development.

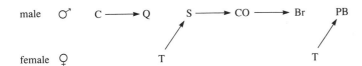

Figure 2.7 The causation of male actions during spermatophore deposition by male newts.

This example illustrates a very important general principle in the causal analysis of behaviour. It may be observed that two behavioural acts, such as Q and S, are strongly associated in time, or temporally correlated, but it is wrong to infer that Q is the cause of S: *correlation does not prove causation*. As an analogy, one may observe that the chimes of two nearby church clocks always strike within a few seconds of one another. The clock that always strikes first could be causing the second to strike, or they could be working independently and slightly out of phase. It is possible to determine which hypothesis is correct by means of an experiment; in this case, one would stop the first clock and see if the second still struck at the expected time. Experiments are critical to the analysis of causation and they will be considered in more detail in Section 2.5.

2.4.3 The functional organization of behaviour

Most sequences of behaviour are fulfilling some obvious biological function and lead towards some clear end-point. Hunting behaviour typically ends with feeding, courtship ends with mating, nest-building ends with the construction of a nest. As a result, behaviour patterns give the appearance of being organized in a functional sense and of being directed towards a specific goal. When a spider builds a web, for example, it follows a number of predictable steps; indeed, if it did not do so, the web would turn out a sorry mess. It first spins a series of radial threads that converge at what eventually will be the centre of the web and then spins a number of tangential threads across them. Such a pattern could be causally organized in one of two ways. It could be the result of a fixed series of 'instructions' from the brain, such as 'spin x radials and then start spinning tangentials', or it could be that the completion of each phase provides the stimuli that elicit the next phase. In the case of a spider's web, it can be quite easily shown that the second hypothesis is correct. If an experimenter removes some of the radial threads at an early stage in web-construction, the spider spins new ones to replace them; it does not proceed to the next phase until the first is complete.

The sequences of behaviour involved in the building of nests by birds can be even more complex. Those built by male weaverbirds, for example, involve intricate basketwork to produce a nest that has a characteristic shape that prevents

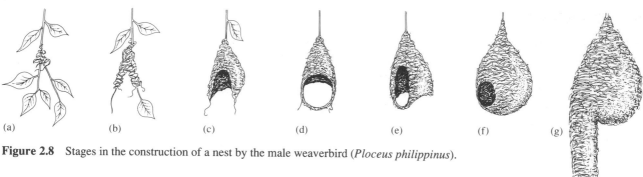

Figure 2.8 Stages in the construction of a nest by the male weaverbird (*Ploceus philippinus*).

predators such as snakes from gaining access to the nest to eat the eggs or chicks (Figure 2.8). Here the hypothesis that males' building behaviour is controlled by a fixed programme is refuted by a number of observations. First, some male weaverbirds are polygamous (they have several mates) and build more than one nest; at any one time, a male may be working on several nests, each at a different stage of construction. Second, if a nest is damaged during construction, a male repairs the damage, often exactly replicating the original structure, before proceeding to the next stage.

A behaviour pattern such as nest building is explicable in terms of each step being a response to the current state of the incomplete nest, so that a bird simply responds appropriately to the state of its nest at a given time. Alternatively, its apparently purposeful behaviour may be 'goal-directed' in the sense that the bird has some concept of what it is trying to achieve. In practice, it is very difficult to tell if behaviour is goal-directed or not, simply because one cannot directly observe what is going on 'inside the bird's head'. Goal-direction is a concept that will be discussed more fully in Chapter 7.

2.5 The experimental analysis of causation

Lack's experiments with robins, described in Section 2.3, showed that, when an animal directs a behavioural response—in this instance attack—towards a stimulus object, only certain aspects of the stimulus object are necessary to elicit the response. A bunch of red feathers would elicit an attack; red feathers are said to be an effective stimulus or a **sign stimulus**, other features of the object being irrelevant. Much of the experimental work carried out in the causal analysis of behaviour is designed to determine those features of objects towards which behaviour is directed that are effective stimuli and those that are not.

Niko Tinbergen carried out a series of experiments on sticklebacks to determine the effective stimuli that elicit attack in males. He found that dead fish that lacked the red belly coloration characteristic of males in the breeding condition did not elicit attack, whereas crude dummies did, whatever their shape, provided that they were painted red on the ventral (lower) side (Figure 2.9). Dummies painted red on the dorsal (upper) side did not, however, elicit aggression. Incidentally, it has since been shown that sexually receptive female sticklebacks also respond to models that are ventrally coloured red by showing a characteristic head-raised receptive posture (Figure 2.12, Section 2.6).

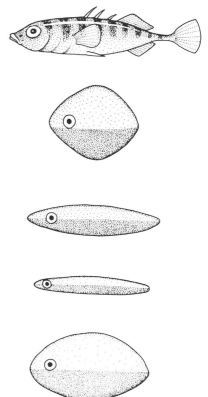

Figure 2.9 Models of sticklebacks that elicit attack in a male. A dead stickleback lacking the red ventral coloration (top) does not elicit attack but a variety of crude models do, provided that they have red colour, shown here as darker shading on the ventral side.

2.5.1   Complex stimuli

It is typically the case that, when animals respond to one particular aspect of a stimulus object, such as when a stickleback responds to the red of an otherwise unrealistic model, they do not respond as strongly as they do to the complete natural stimulus. This suggests that there may be more than one effective component to a natural stimulus. This idea has been investigated in great detail by G. P. Baerends and J. P. Kruijt (1973), who investigated the retrieval response of herring gulls (*Larus argentatus*) to eggs placed just outside their nest. Like greylag geese, herring gulls will move an egg back into their nest. Baerends and Kruijt set out to investigate the relative importance of four aspects of eggs in eliciting this response: their size, shape, colour and whether or not they are speckled. Herring gulls generally lay three eggs; the basic experiment was to remove two of them from a nest, to place two dummy eggs on the edge of the nest, and to record which of them the gull retrieved first, using this as an indicator of their preference. Before they could use this method to compare eggs of different size, colour or shape, they had to control for an interesting feature of the gulls' behaviour. If two identical dummies are placed on the edge of the nest, an individual gull usually has a strong position preference, retrieving the right before the left, or *vice versa*. This effect was countered by balancing the position effect against the effect of size. For example, an individual bird with a right-egg preference was given a series of tests in which eggs of different size were presented, with the larger always on the left, the non-preferred side. When the size difference between the two eggs is such that the gull is equally likely to retrieve either of them, the position effect has been cancelled out. The ratio of the sizes of the two eggs provides a numerical measure of that individual gull's position preference.

A large number of experiments of this kind yielded the following conclusions, summarized in Figure 2.10 (*overleaf*):

1   When dummies are brown and speckled, like real eggs, gulls preferred larger to smaller eggs. They even preferred eggs twice as large as natural eggs.

2   Green eggs were preferred to those of a natural brown colour.

3   Speckled eggs were preferred to plain eggs.

4   Egg-shaped dummies were preferred to block-shaped dummies.

Thus, whatever the shape and colour of a particular series of dummies, larger ones are always preferred to smaller ones. The various features tested have an additive effect on the strength of a gull's response to a specific dummy. This phenomenon is called **heterogeneous summation**, meaning that the various features of a particular stimulus are additive in their effectiveness in eliciting a response.

The dummy that most closely resembles a real egg in Figure 2.10 is that of size 8 in row R. There is thus a great variety of objects, particularly those that are larger and green, that are preferred to the natural egg dummy. A stimulus that is more effective at eliciting a response than the natural stimulus for that response is called a **supernormal stimulus**.

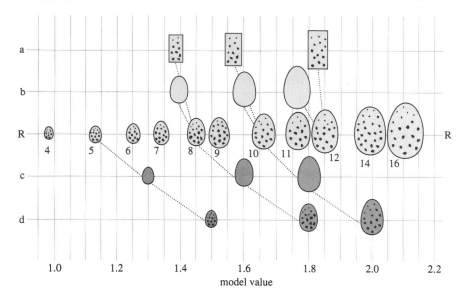

Figure 2.10 The preferences of herring gulls for objects varying in size, shape and colour. Within each row, preference increases from left to right. Row R shows a series of normally-coloured (brown and speckled), egg-shaped dummies that varied in size (a normal herring gull's egg has a value of about 8). Row a are brown, speckled, rectangular blocks; row b are brown, unspeckled, egg-shaped; row c are green, unspeckled, egg-shaped; row d are green, speckled egg-shaped. Dotted lines join objects of comparable size.

Retrieval of an egg just outside its own nest is only one of a number of responses that a herring gull may direct towards an egg-like object. If the egg is inside the nest, the gull sits on it and incubates it; if it finds it some way away from its own nest it eats it. Experiments similar to those described above have shown that each kind of response is elicited by different specific features of eggs. The features relevant to retrieval are colour, speckling and size; shape is relatively unimportant. Objects must be egg-shaped to elicit predation, and they must have rounded edges for a bird to settle on them and incubate them. Thus, these three different activities of retrieval, predation and incubation are elicited by different aspects of the same stimulus object. This makes two points of general importance. First, different aspects of a stimulus object are relevant in different contexts; thus, the particular aspects to which an animal responds at a given time depends on the context—what the animal is doing, or where it is—at that time. Second, the fact that one feature is important in one context but not in another rules out the possibility that that feature is unimportant in any one context because the animal cannot detect it. It appears that animals such as herring gulls perceive many aspects of a stimulus object but, in a given context, attend selectively to particular features of that object and pay less attention to others.

2.5.2 Sensory exploitation

Where patterns of behaviour are elicited by specific sign stimuli, this provides an efficient and reliable means by which an animal can respond appropriately to its environment. It can, however, open the way for animals of another species to exploit a stimulus–response relationship to their own advantage. For example, cuckoos lay their eggs in the nests of other birds, usually much smaller than themselves. The larger egg of the cuckoo acts as a supernormal stimulus to the foster parents, which incubate it and, when it hatches, the huge gaping mouth of the cuckoo chick begging for food acts as a supernormal stimulus that causes them to devote all their time and energy to feeding it.

Perhaps the most remarkable example of the exploitation of one species' stimulus–response system by another species is provided by certain fireflies.

Males of the genus *Photinus* fly about at night, emitting a species-specific pattern of flashes, to which receptive females of the same species, sitting near the ground, reply with a distinctive flash pattern of their own. Male and female signal to each other until the male lands beside the female and they mate. Females of another firefly genus, *Photuris*, mimic the flash patterns of female *Photinus* and thus lure *Photinus* males to the ground where they devour them. Females of some species showing this '*femme fatale*' behaviour can mimic the female flash patterns of several prey species.

2.5.3   Stimuli and orientation

Features of the environment can profoundly influence the behaviour of animals without acting as releasers of specific patterns of behaviour. They may, for example, serve as landmarks that enable animals to find their way around their home range and to guide their movements. This aspect of behaviour was investigated experimentally in a classic study by Niko Tinbergen of the digger wasp (*Philanthus triangulum*). The female digger wasp digs a burrow in sand in which she lays her eggs and then flies off to catch insect prey that she brings back to her burrow for the larvae that hatch from her eggs to feed on. Tinbergen waited until a female wasp was in her burrow and then placed a ring of pine cones around the entrance (Figure 2.11). When the female emerged she flew around the burrow entrance a few times before flying off in search of prey. While she was away, Tinbergen moved the ring of pine cones a few inches to one side of the nest entrance. On her return, the wasp flew directly to the centre of the ring, thus missing the entrance to her burrow. This simple experiment shows that the female used the ring of pine cones as stimuli to guide her return to her nest entrance.

The question of how animals perceive their environment, and how their behaviour is guided by their perception of it, will be discussed further in Section 2.7.

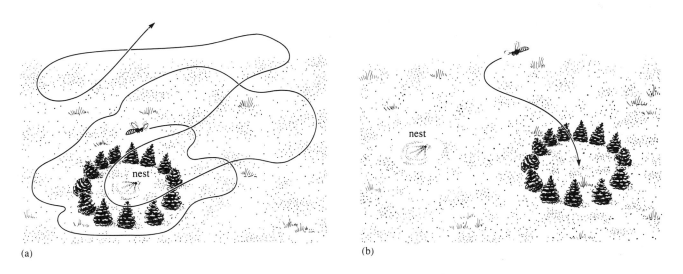

(a) (b)

Figure 2.11   Experimental manipulation of orientation by the female digger wasp. *Left*: a ring of pine cones is placed around the entrance to a female's burrow while she is inside it. When she emerges, she flies around the unfamiliar objects a few times before flying off in search of prey. *Right*: during her absence, the ring of pine cones is moved to one side of the burrow. On her return, the female digger wasp goes directly to the centre of the ring of cones.

2.6   Stimulus–response chains

Many complex patterns of social behaviour consist of a number of stimulus–response relationships, in which a specific action by one animal acts as a stimulus eliciting a specific response in another. A good example is provided by the courtship behaviour of the three-spined stickleback (*Gasterosteus aculeatus*) (Figure 2.12). When a gravid female (gravid means full of ripe eggs) enters a male's territory, he performs a characteristic swimming pattern, called a zigzag dance, in which he swims rapidly back and forth across his territory. If the female is receptive (i.e. willing to mate), she adopts a head-up posture. The male then swims down to his nest and the female follows him. At the nest, the male repeatedly thrusts his snout into the entrance (showing the nest) and the female enters it. The male then trembles his snout against the base of her tail and she lays

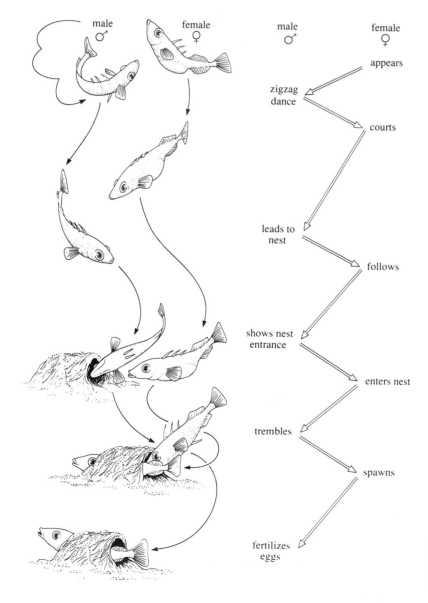

Figure 2.12   The courtship sequence of the three-spined stickleback represented as a stimulus–response chain.

her eggs in the nest (spawning). The female then swims out of the nest, through the end opposite to the entrance, and the male follows her through, ejaculating sperm onto the eggs as he does so. The male usually then chases the female out of his territory.

By making repeated observations of courting sticklebacks, Tinbergen established sequential correlations between the various elements in the sequence and concluded that each action by one partner was both a response to what the other fish had just done and was also the stimulus that elicited the next behaviour by the partner. The sequence could thus be depicted as a chain of stimuli and responses, in which each behavioural element was both a response and a stimulus (Figure 2.12). In fact, the behaviour of sticklebacks is not as stereotyped as this; following a particular action by one fish, the partner may show one of a number of possible responses. A more realistic picture, based on a study of the courtship of the ten-spined stickleback (*Pungitius pungitius*) by Desmond Morris, is shown in Figure 2.13.

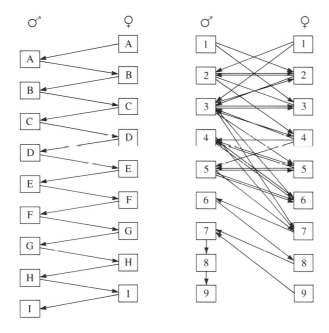

Figure 2.13 The courtship sequence of the ten-spined stickleback represented as a stimulus–response chain. *Left:* the idealized sequence. *Right:* the actual pattern of responses observed.

The actual sequence of behaviour is much more variable than that depicted in the idealized stimulus–response chain, but it is clear that the behaviour of each sex is not random with respect to the other. Thus, given that one fish has performed a particular action, what its partner will do next can sometimes be predicted with reasonable accuracy. For example, male action 6 is invariably followed by female action 8, which invariably precedes male action 7; male action 3, however, can be followed by female actions 2, 3, 4, 5, 6 or 7. In general, actions by one sex are closely-linked, in numerical terms, to actions by the partner, so that there are very few instances where either fish responds with an activity that is more than two stages ahead of that performed by its partner.

☐ To what do you suppose this variation in the response of one fish to the behaviour of the other is attributable?

■ To motivational effects. It is a reasonable hypothesis that, where one partner responds to an early action by its partner by performing a much later action that it is more strongly motivated to mate than it is when it responds with an earlier action in the courtship sequence.

2.6.1 Consummatory stimuli

At the end of a sequence of behaviour, an animal will frequently stop performing that general category of behaviour and switch to another activity. At the end of the stickleback courtship sequence, for example, the male attacks the female, drives her out of his territory and adopts other activities, such as tending his nest, feeding and defending his territory. What causes him to switch from one activity to another? One possibility is that there is an external stimulus to which the male's response is to change his behaviour. Such a stimulus is called a **consummatory stimulus**.

Factors that bring about the end of three-spined stickleback courtship were investigated in great detail by Sevenster-Bol (1962). Specifically, she focussed on the observation that the ratio of zigzag movements (courtship) to attacks (aggression) shown by a male is high at the start of courtship and is very low immediately after courtship. She considered three possible factors that might bring about this change:

1 The performance of courtship behaviour.

2 The act of ejaculation.

3 The presence of a clutch of eggs in the nest.

Sevenster-Bol performed a number of experiments to separate these three effects. For example, males could be exposed to factor 3, but not 1 and 2, by experimentally placing a clutch of eggs in a male's nest. It was found that whether or not courtship occurred had very little effect on a male's behaviour. The results from tests in which fertilization and the presence of a clutch were manipulated are shown in Figure 2.14.

Consider the data presented in Figure 2.14.

☐ How, in general, does the male's behaviour change after courtship?

■ The number of zigzags performed per test increases with each successive test following the end of courtship.

☐ Is the male's behaviour affected by whether or not he performs fertilization?

■ No. From Figure 2.14a it is clear that the increase in zigzagging follows a similar pattern in each condition.

☐ Is the male's behaviour affected by whether or not he has a clutch of eggs in his nest?

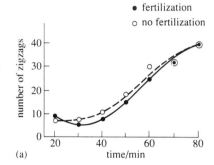

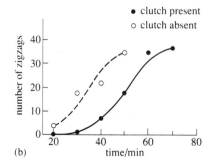

Figure 2.14 Changes in the number of zigzag movements performed by male three-spined sticklebacks to females in a series of tests, conducted at 10-minute intervals after the end of courtship, according to whether or not eggs were present in the male's nest and whether or not fertilization occurred. (a) Clutch always present, fertilization varied. (b) Fertilization always occurred, presence of clutch varied.

■   Yes. Figure 2.14b shows that the frequency of zigzagging increases much more rapidly over successive tests when eggs are absent than when they are present.

☐   In what sense, then, is the presence of eggs in the nest a consummatory stimulus?

■   The presence of eggs in the nest appears to inhibit, or to delay the increase of zigzag behaviour. In other words, it inhibits further courtship.

2.7   The sensory guidance of behaviour

This section returns to an aspect of behaviour that was introduced at the end of Section 2.3, the way that an animal's behaviour is guided by objects and events in its external world. There are two general questions:

1   How do animals use their senses to guide their behaviour?

2   How do the sensory systems of animals extract the sensory information that is necessary to guide their behaviour?

2.7.1   Orientation in space: taxis and kinesis

Animals move about in an environment that is spatially heterogeneous, meaning that they will survive better in some places than in others. Some places offer a richer food supply, or better protection from predation than others, for example. How do their senses enable them to find these better places? Some simple principles can be derived from the study of invertebrate animals, in which the sensory guidance of movement can sometimes be very easily studied. A basic distinction can be made between two kinds of movement: **taxis** and **kinesis**. Taxis refers to movement that is directed with respect to the location or source of a stimulus. Kinesis refers to movement in which the animal responds to an environmental stimulus but does not orient with respect to the location or source of that stimulus. The following examples of some specific patterns of behaviour illustrate just some of the ways by which animals find their way around their environment (orientation), and will make this distinction clear.

1   Woodlice are creatures that aggregate in damp places; if they are exposed for too long to dry conditions, they die of desiccation. A woodlouse cannot detect damp places from a distance. If it is placed in a damp environment it moves around very little and rather slowly, but, when placed in a dry environment, its locomotor activity and the speed of its movements increase. As a result, woodlice in damp places will tend to remain there, those in dry places will tend to move away from them, with a reasonable probability that they will encounter damper conditions.

2   A housefly maggot will move away from a source of bright light. It makes repeated turns of its head, on which there are light-sensitive eyes, from side to side. It turns its body away from the side on which it receives stronger light stimulation.

3   The pill woodlouse (the kind that rolls itself into a ball when disturbed) moves towards light if it has been kept dry and deprived of food. It has two

eyes and so could, potentially, detect a source of light by comparing the amount of light entering each eye. That the pill woodlouse does this is supported by the observation that, if one eye is covered up, the animal moves in a circle.

4   Ants follow definite paths between their nests and sources of food and experiments have shown that they do this by moving along in a direction at a certain angle to the position of the sun. This can be demonstrated by altering the apparent position of the sun through a certain angle with a mirror, which results in the ants moving along a path that is displaced by a similar angle. Thus, they do not move directly towards or away from the stimulus that guides their movement, but use it as an appropriate reference point.

☐   Which of these patterns of behaviour (1–4) can be classed as examples of taxis, which as examples of kinesis?

■   Example 1 is a kinesis, 2–4 are examples of taxis.

Consider again the behaviour described in example 4.

☐   What problem does any animal, including humans, face when using the sun as a direction reference?

■   The sun moves across the sky during the day.

If ants always moved in a direction that was, say 45° to the sun, their path would shift during the day and they would fail to get to their destination. In fact, like many animals, ants compensate for the sun's movements, an ability that requires that they have a sense of time or 'internal clock'. We humans can use the sun as a compass if we know the time; if we know that it is midday, for example, south is where the sun is (as long as we are north of the equator). As described later in the course (Book 5), an accurate time-sense is present in many animals.

Orientation with reference to specific stimuli

Many moths locate mating partners by means of airborne odours, called **pheromones**. (A pheromone is a chemical substance produced by an individual animal that alters the behaviour or physiology of another individual.) In many species, the female remains stationary and secretes pheromone which is carried away downwind in what is called an 'odour plume', in which the concentration of pheromone is higher closer to the moth than far away (Figure 2.15). Males find females by flying up the odour plume; if they fly out of it, and cease to receive sensory input, they take appropriate action and change course to fly back into the plume. Snails use visual information to find their way around. Given a choice of identical long, thin objects, they will crawl preferentially towards those that are placed vertically to those that are placed horizontally. Thin, vertical objects approximate visually to the stems of plants on which snails feed. In this case, the snail appears to perceive and respond appropriately to specific patterns in its visual field, a topic discussed in more detail in the next section.

When an animal is presented with a stimulus, the exact nature of its response to it may be influenced by several factors. This is illustrated by the jumping response of frogs when threatened by an approaching large object (Figure 2.16). Frogs respond to such stimuli by jumping. In terms of maximizing the distance achieved in a jump, the frog does best, for purely mechanical reasons, if it jumps directly forwards. This is not necessarily the best direction to jump, however, in terms of getting away from the threatening object. In addition, the frog may have to jump in another direction if there is an obstacle in the way. The actual direction in which it jumps can be seen as a compromise between these conflicting components of the situation.

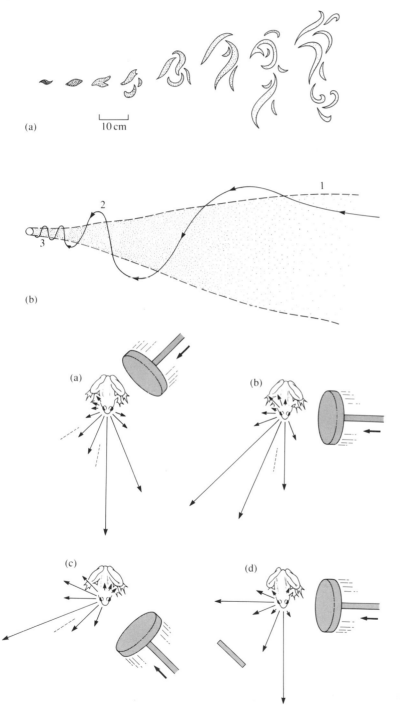

Figure 2.15   Olfactory communication in insects. (a) Artist's conception of an odour plume; note that the odour is produced in a series of discrete pulses. (b) Hypothetical path of an insect locating an odour source by:
(1) orienting upwind in an area of low odour concentration; (2) flying out of the plume, losing the trail, and then using crosswind flight to relocate it;
(3) entering an area of high odour concentration where the odour source can be located visually.

Figure 2.16   Direction of avoidance jumps by the frog *Rana pipiens* when threatened by an approaching object. The length of the arrows radiating from the frog's head represents the frequency distribution of different jump directions (the longer the arrow, the more frequent are jumps in that direction). (a) to (c) show that the preferred jump direction is a compromise between jumping directly forwards and jumping directly away from the object. (d) shows that the frog adopts other directions to avoid an obstacle placed in the path of its normally preferred jump direction.

2.7.2 Stimulus filtering

Human eyes and ears are responsive to a wide range of wavelengths of light and sound, respectively; they are 'broad band' detectors. Many other species have similar receptors, though the range of frequencies to which they are responsive may be different. Insects, for example, can see a wider range of light wavelengths than ourselves, extending into the ultraviolet, as described in Section 2.3.1. Snakes have extended sensitivity in the other direction, into the infrared. They do not perceive infrared light with their eyes, but with specialized pit-like organs that resemble pinhole cameras, situated between the eyes and the nostrils (Figure 2.17). They use this sense to detect and locate their warm-blooded prey, such as rats and mice, which emit infrared radiation.

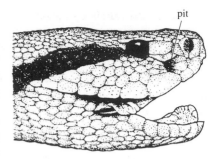

Figure 2.17 The head of a rattlesnake, showing the pit that is sensitive to infrared radiation.

Many of the sense organs used by other animals are not broad-band receptors but are responsive to a very narrow band of frequencies. They are finely-tuned to particular frequencies of light or sound, or to particular odours, that are of special biological relevance to the animal. For example, the female silk moth (*Bombyx mori*) produces a volatile substance called bombykol that she emits as an odour plume like that shown in Figure 2.15. The male silk moth possesses huge antennae that carry many thousands of receptor cells that are responsive only to bombykol. Such is the sensitivity of these receptors that only one or two molecules of bombykol are sufficient to cause a receptor to respond, enabling a male to detect a female from several kilometres away.

Some sense organs are 'broad-band' receptors but contain sensory mechanisms that respond selectively to particular features of the external world. Frogs and toads, for example, have large eyes that enable them to see their general environment, and particular objects in it, such as mates and enemies. Their eyes also contain specialized receptors that see small insects; these so-called 'bug-detectors' have been studied in some detail. A small electrode can be placed in an appropriate part of a frog's visual system and the frog can be presented with visual stimuli on a screen (Figure 2.18). The images that the frog sees can be varied, and its sensitivity to different visual stimuli can be measured by recording nervous activity through the electrode. If the frog is presented with a field of small dots that are stationary or that all move together, no response occurs (Figure 2.18a). If, however, just one of the dots is moving, the frog's nervous system responds. This effect is more marked if the dot moves in a irregular fashion against a complex background. Thus, the more the stimulus resembles a real insect, the stronger the response.

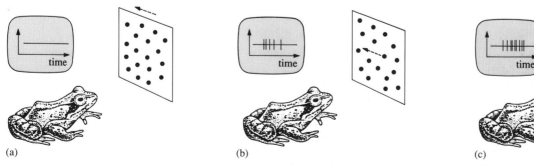

(a) (b) (c)

Figure 2.18 The responsiveness of a frog's nervous system, measured as electrical activity (as in Figure 2.5), to (a) a field of dots that are stationary or all moving together; (b) a field of dots in which only one dot is moving; (c) a moving insect against a complex background.

The stimuli that elicit prey-catching and predator-avoidance in common toads (*Bufo bufo*) have been studied extensively by J-P. Ewert and his colleagues. Toads show anti-predator avoidance responses (turning and jumping away) to large, moving objects that approach them from above; they show feeding responses (turning, tracking and snapping) to small, moving objects at ground level. These feeding responses were examined in detail by presenting toads with simplified stimuli, consisting of black rectangles moving horizontally across a white background. They responded most strongly to an elongated rectangle moving in the direction of its long axis, the so-called worm configuration (Figure 2.19). An identical rectangle moving in the direction of its short axis (the anti-worm configuration) elicits very little response. The response given to a black square is intermediate, suggesting that the stimulating effect of a horizontal long edge moving horizontally and the inhibitory effect of a vertical long edge moving horizontally are additive.

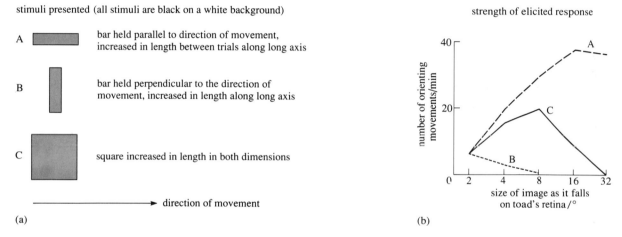

(a)

(b)

Figure 2.19 The prey-catching response of toads to various stimuli moving across their visual field. (a) Test stimuli. (b) Responses to test stimuli.

In these examples, animals respond selectively to particular kinds of stimulus, because the nature of their sense organs is such that they are particularly responsive to those stimuli. This kind of stimulus filtering, based on the properties of sense organs, is called **peripheral filtering**. It is differentiated from stimulus filtering that is due to processes occurring, not in the sense organs, but in more central parts of the nervous system, called central processing. One kind of central processing, called selective attention, is familiar in everyday life. For example, when you are at a party you are conscious that there is much conversation around you, though you may be attending to what a particular individual is saying to you, and not be aware of what other people in the room are saying. Should someone mention your name, however, you are typically immediately aware of the fact; somehow your sensory system is particularly alert to the sound of your own name. Many parents of small babies are able to sleep through quite loud noises but will wake up immediately should their baby make the slightest noise. It appears that human ears are gathering a great deal of auditory information all the time, but that, somewhere in the brain, the information is being processed and scanned for stimuli of immediate relevance.

A form of selective attention that has been extensively studied by biologists involves an apparent change in perception, shown by animals searching for food. Animals that use vision to find their food, such as birds, commonly have to find food that is highly cryptic, that is, well-camouflaged against its background. Field observations of a variety of birds indicate that, having found one item of a particular food type, a bird will then eat several more items of that type, before switching to another type. This suggested that the birds form a search image, meaning that, having found one item they subsequently found that type of food easier to see. A thorough investigation of the search image hypothesis was carried out on domestic chicks (*Gallus gallus domesticus*) by Marian Dawkins (1971). She first fed chicks for three weeks on either orange or green coloured rice grains presented on a white background. They were then tested for their ability to detect the coloured rice grains on a background of the opposite colour, on which the grains were conspicuous, and on a background of the same colour, on which they were cryptic (Figure 2.20). The results are shown in Figure 2.21.

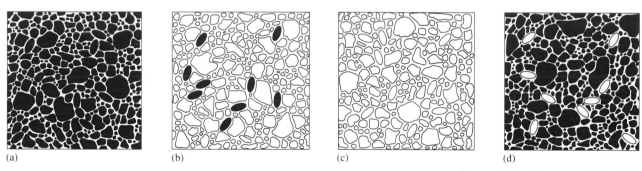

(a)   (b)   (c)   (d)

Figure 2.20 Rice presented to chicks on different backgrounds: (a) green rice on a green background; (b) green rice on an orange background; (c) orange rice on an orange background; (d) orange rice on a green background.

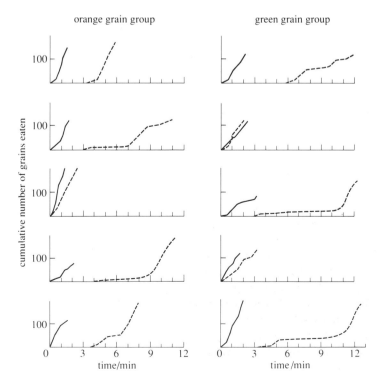

orange grain group   green grain group

cumulative number of grains eaten

0   3   6   9   12
time/min

0   3   6   9   12
time/min

Figure 2.21 Individual records for 10 chicks, showing the cumulative number of rice grains eaten in successive periods of 15 s. Solid lines: conspicuous grains; dotted lines: cryptic grains.

Dawkins found that individual chicks varied considerably in their behaviour, but that certain general effects could be detected.

☐   How did the 10 chicks typically behave when feeding on conspicuous rice grains?

■   They all found the food very quickly at the start of the test and then ate it at a high rate, as shown by the steepness of the cumulative intake plots.

☐   How did the chicks behave when feeding on cryptic rice grains?

■   A few of the chicks found the cryptic food very quickly, but most did not start to find it for several minutes. Once they began to find the cryptic grains, most of the chicks found them at an increasing rate, as shown by the increasing slope of the cumulative intake plots.

Expressing some of these effects in figures, the mean time taken by the chicks to eat the first five rice grains was 9.1 s in the conspicuous tests, 240.0 s in the cryptic tests; in contrast, the time taken to eat the last five grains (the 96th to 100th grains) was 3.0 s in the conspicuous grain tests, 3.6 s in the cryptic grain tests.

☐   Do these results support the hypothesis that the chicks form a search image for cryptic food?

■   Yes, they do. It is clear that most of the chicks were initially slow to find the cryptic prey, but that, by the end of a trial, they could find them as fast as they could find conspicuous grains.

It is not clear, in terms of what is going on in the nervous systyem, what the formation of a search image actually involves. There is clearly a change in the way the animal perceives its external environment; what was undetectable to it, becomes detectable. Inasmuch as the experience of finding one cryptic food item alters a chick's ability to detect other, similar items, the formation of a search image can be regarded as a learning process; essentially the chick 'learns to see' its food. This raises an important point. When it is found that an animal possesses the ability to respond selectively to some specific aspect of its sensory world, it should not be assumed that this is explicable in terms of the neurophysiology of its sense organs, as is the ability of frogs to detect insects. The ability may be the result of learning earlier in the animal's life and may involve changes that are much more central within the nervous system.

2.8   Senses and communication

Behaviour sequences that involve communication—an interaction between animals—are a particularly fruitful source of examples of stimulus–response systems. In many instances, there is a very high degree of specificity in the relationship between a signal produced by one animal and the response shown by another. This is especially true of communication between males and females during courtship and mating. Male frogs and insects produce calls that are very stereotyped and which differ markedly from one species to another; their signals are said to be species-specific. Among fireflies, males produce species-specific patterns of flashes (Figure 2.22). The function of this species-specificity is that it enables females to recognize and respond only to males of their own species and to avoid mating with males of another species. (Mating with members of another species, called hybridization, typically produces young that either fail to develop or that are less likely to survive and breed than pure-bred offspring.) In many species, the sense organs of females are specifically receptive to the signals produced by **conspecific** (of the same species) males. The ears of female frogs, for example, contain a large number of receptors that are sensitive to the particular sound frequencies contained in the calls of conspecific males, and very few that are sensitive to the sounds produced by heterospecific (of other species) males. Their ears are said to be 'tuned' to the sounds of conspecific males and they are, in effect, deaf to the sounds of heterospecific males.

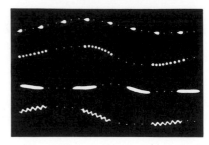

Figure 2.22   The flash patterns made by male fireflies of four different *Photinus* species as they fly through the air from left to right.

The specificity of signals is also often influenced by the nature of the environment in which the animal is signalling. For example, fireflies that signal at night, in full darkness, produce green light, which is particularly visible in the dark. Some firefly species signal and mate at dusk, however, when, because of the reflection of light from foliage, there is a lot of ambient green light. A survey of fireflies active at dusk showed that, of 23 species examined, 21 produced yellow rather than green light. Yellow light is much more visible at dusk than green light.

It is by no means universal in sexual signalling systems that it is the male that signals and the female that responds. We have already seen one example, the silk moth, in which the reverse is true; the female produces pheromone and the male follows an odour plume to find her.

The specificity of male signals, and of female response to conspecific signals, is the result of natural selection acting separately on the two sexes. Selection favours males that give the kind of signal that is most easily detected and recognized by conspecific females. It favours females that possess sense organs that can detect and recognize those signals. This process, in which selection on one sex is accompanied by a different but complementary selection on the other sex is called **coevolution** between the sexes. In some species, coevolution between the sexes has led to some remarkable examples of signal and receiver specificity.

In an African fish, *Haplochromis burtoni*, the female broods her fertilized eggs in her mouth; she swims around with them in her mouth (with her tendency to feed suppressed!) until they hatch, when she opens her mouth to release the fry. During courtship and mating the eggs have to be laid, fertilized by the male and then taken up into the female's mouth. At the end of a sequence of courtship acts, the female sheds her eggs onto a rock and the male ejaculates sperm onto them. The female immediately starts to suck up the eggs. The male has a row of coloured spots along the edge of his ventral fin that bear a striking resemblance to the eggs (Figure 2.23). As she sucks up her eggs, she also attempts to suck up these egg-spots; because they are close to the male's genital opening she also sucks up sperm and takes it into her mouth, where it fertilizes any eggs that are not yet fertilized.

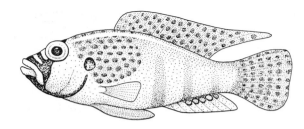

Figure 2.23   A male *Haplochromis burtoni*, showing the row of egg-spots on the ventral fin.

2.9   Hormones and the causation of behaviour

Many of the responses that animals show to external stimuli show variation over time. An animal may be more responsive to a particular stimulus at a particular time of day or at a particular time of year. This is especially true of sexual behaviour, which is commonly a seasonal phenomenon. For example, male birds show sexual responses to females and aggressive responses to males in the breeding season, but respond similarly to the two sexes at other times of year. These variations often reflect changes in the animal's internal state due to the action of hormones. A **hormone** (from a Greek word meaning 'arouse to activity') is a chemical compound that is secreted within the body by an **endocrine gland** and released into the bloodstream. Although it is thus carried to all parts of the body in the blood, it does not necessarily affect all the tissues it reaches. Some cells of the body will be responsive to it while others will not; only those cells with appropriate receptors will respond to a particular hormone.

Hormones are diverse in terms of their chemical composition and they show a huge diversity of behavioural effects; it is possible here only to establish some general principles. Hormones are themselves secreted in response to external and internal stimuli. Sex hormones such as **testosterone** are commonly secreted in response to changes in daylength, for example (Figure 2.24, *overleaf*). The sex hormones are hormones that are produced by and that influence the behaviour of both sexes. They are mainly secreted by the *gonads*, the sex organs that produce gametes—ovaries in females and testes in males.

☐   Study Figure 2.24. What major changes in the birds' behaviour are indicated by the first three lines?

■   From being highly sociable in the winter, flocking and roosting together, starlings become aggressive and territorial, and show a lot of sexual behaviour in the spring.

Hormones frequently act together with external stimuli to elicit behaviour. For example, the presence of a female will elicit courtship responses in a male stickleback or newt only in the spring, when males have high levels of

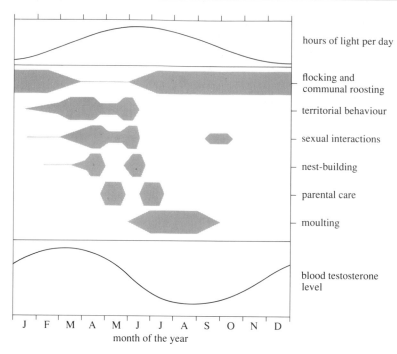

Figure 2.24 Events in the annual reproductive cycle of starlings (*Sturnus vulgaris*). As the number of hours of daylight increase in the spring (top line), the level of testosterone in the blood reaches a peak (bottom line). This affects the amount of time (indicated by the depth of the horizontal bars) that the birds devote to various activities.

testosterone in their blood. The heightened level of testosterone is itself a response to an external change, called a *priming* stimulus, in this instance an increase in daylength (Figure 2.25).

Unlike many of the external stimuli discussed in previous sections, which generally have a very specific effect on behaviour, a single hormone can have several effects of different kinds in different parts of the body and may influence several aspects of behaviour, often relating to very different activities. Testosterone, for example, influences not only sexual behaviour but also aggression, and there is evidence, from chicks and mice, that it increases the persistence with which individuals search for food.

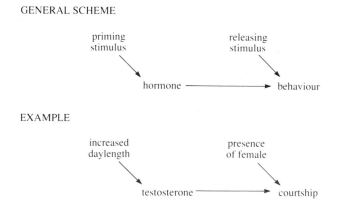

Figure 2.25 The interaction between external stimuli, hormones and behaviour.

Three general categories of hormonal effect have been identified: *central effects*, acting on the brain; *peripheral effects*, acting on a particular part of the body other than the brain; and *signal effects*, causing the animal to present different stimuli to other animals. For example, the hormone **oestrogen** acts on the brain of a female cat (a central effect), causing it to secrete other hormones that affect her ovaries and other reproductive organs. It also has a direct effect on her vagina, making it highly sensitive to touch stimuli (a peripheral effect). Finally, it influences glands in her genital region, causing them to emit pheromones (a signal effect), which are attractive to males.

There is a widespread popular misapprehension that hormones are only involved in sex and reproduction. This is far from true, as the following examples show. The hormone insulin, together with another hormone called glucagon, controls the level of glucose in the blood; inadequate amounts of insulin lead to the condition of diabetes. **Adrenalin** is a fast-acting hormone, secreted at times of acute stress. It is sometimes called the 'fight or flight' hormone, and it brings about the effects that one feels when suddenly angered or alarmed, such as a faster heart-rate and increased alertness. Growth hormone has a slow action and affects the formation and growth of tissues such as bone; low levels in humans causes dwarfism.

Hormones often influence behaviour over a relatively long time-scale. High levels of sex hormones for example, are generally necessary for the expression of sexual behaviour, but levels of these hormones typically change rather slowly and so minute-to-minute changes in sexual behaviour cannot be attributed to concurrent changes in hormone levels but must be explained as responses to more immediate events, such as the behaviour of the mate. The time-scale over which most hormones act on behaviour may be an animal's lifetime, as in sexual maturation (puberty in mammals), a year, as in animals that breed annually, or a matter of days or weeks, as in the control of reproductive cycles in mammals. The hormone adrenalin is rather unusual in having an effect on behaviour over a matter of seconds or minutes, though current research suggests that there are more such hormones to be discovered.

Summary of Chapter 2

Much of the behaviour of animals can be understood, in causal terms, as responses to objects and events in their external environment. This is most apparent in the case of reflex responses, such as the knee-jerk response, but most responses to external stimuli show considerable variation. The nature of the stimuli to which a particular animal will respond is largely determined by the information that it receives through its sense organs. Animals do not have complete information about their external environment; each species has a range of sense organs that provide it with a partial and selective representation of its external world. In addition to information from the exterior, provided by exteroceptors, the nervous systems of animals receive information about the disposition of parts of their body (from proprioceptors) and about their internal state (from enteroceptors). If an animal is to respond appropriately and accurately to external stimuli, it must integrate exteroceptive and proprioceptive information (reafference).

An animal can respond in several ways to a particular stimulus. Not only can different animals respond in different ways to the same stimulus, but an individual animal can respond to the same stimulus in different ways at different times.

Many behaviour patterns are highly stereotyped and species-specific. These features are central to Lorenz's concept of a fixed action pattern (FAP), which he saw as being inherited and controlled by fixed programs in the nervous system. Some patterns of behaviour, such as the escape response of the sea slug *Tritonia*, fit this model quite well, but most do not. For most behaviour, it provides an inadequate analysis of causal mechanisms. A major reason why this is so is that responses to specific stimuli are generally not stereotyped but show variation. Much of this variation is due to the animal's internal state at the time, what is referred to as its motivation. This aspect of behaviour is examined in detail in Chapter 7 of this book.

Behaviour patterns, especially those involving interactions between one individual and another, are commonly organized into more or less stereotyped sequences of behaviour. Within a sequence, a specific behaviour pattern may be causally related either to an external stimulus, or to the animal's preceding behaviour. In addition, the precise sequence of actions within a sequence may be influenced by feedback from the consequences of those actions, notably in the case of consummatory stimuli that cause sequences to terminate.

An external stimulus may present several distinct features to an animal, such as its size, shape and colour. The animal may respond selectively to one or more of these features. Variation in the various features may have an additive effect, called heterogeneous summation. In some instances, artificial or natural stimuli (called supernormal stimuli) have been identified which are more effective at eliciting a response than normal stimuli.

As it moves around its environment, an animal responds to external objects. These may serve as stimuli that guide its behaviour in general, rather than eliciting specific responses. This may take the form of examples of taxis or kinesis, or may involve more complex forms of spatial orientation, as in the use of objects as landmarks.

The sensory systems of animals are selective, 'filtering out' information that is biologically relevant. This filtering may take place peripherally, in the sense organs themselves, or may occur more centrally in the nervous system, as in selective attention and the formation of search images.

Communication between animals, in which behaviour by one animal acts as a stimulus eliciting a response in another, involves the coevolution of signalling systems and the sensory systems that receive communicative stimuli.

The responses of animals to particular stimuli may be affected by their hormonal state. Hormones are secreted in response to external stimuli, such as daylength, as well as to internal stimuli.

Objectives for Chapter 2

When you have completed this chapter, you should be able to:

2.1   Define and use, or recognize definitions and applications of each of the terms printed in **bold** in the text. (*Question 2.6*)

2.2   Understand and describe what is meant by a reflex and discuss the extent to which behaviour can and cannot be understood in terms of reflexes.

2.3   Discuss the ways in which animals are more than passive receivers of stimuli from their environment.

2.4   Describe what is meant by reafference and discuss its importance in behaviour. (*Question 2.1*)

2.5   Categorize the different kinds of senses that animals possess. (*Question 2.2*)

2.6   Describe what is meant by a fixed action pattern (FAP) and discuss the ways in which it is of limited use as an explanatory concept in the study of behaviour. (*Question 2.3*)

2.7   Give examples of behaviour patterns that, to varying degrees, conform to the FAP concept. (*Question 2.3*)

2.8   Explain why, in the analysis of sequences of behaviour, correlation is not proof of causation, and give examples of experimental tests of the hypothesis that correlation is evidence of causation. (*Question 2.4*)

2.9   Explain, with examples, what is meant by a sign stimulus. (*Question 2.5*)

2.10   Describe an example of how the relevant components of a stimulus are identified by means of experiments.

2.11   Explain, with examples, what is meant by heterogeneous summation. (*Question 2.5*)

2.12   Describe examples of how animals use objects in their environment to guide their behaviour and describe the role of experiments in this kind of analysis.

2.13   Describe, with examples, what is meant by a stimulus–response chain. (*Question 2.6*)

2.14   Describe, with examples, what is meant by a consummatory stimulus. (*Question 2.6*)

2.15   Differentiate, with examples, between a taxis and a kinesis. (*Question 2.6*)

2.16   Explain, with examples, what is meant by stimulus filtering and differentiate between peripheral and central filtering and central processing. (*Question 2.7*)

2.17   Explain what is meant by selective attention and discuss evidence for a specific example, search image formation.

2.18   Explain, with examples, what is meant by species-specificity in communication.

2.19   Discuss, with examples, the concept of coevoluss with reference to the evolution of communication systems related to mating. (*Question* 2.7)

2.20   Describe how hormones and external stimuli interact in the causation of behaviour.

Questions for Chapter 2

Question 2.1 (*Objective 2.4*)
A lizard approaches and catches an insect crawling across a wall. What features of the situation provide the lizard with (a) exafferent (b) reafferent visual information? How are the two kinds of information used by the lizard to catch the insect?

Question 2.2 (*Objective 2.5*)
What kind of senses—enteroceptive, proprioceptive, or exteroceptive—are you using in the following three instances? (i) You are awoken by an alarm clock in the morning. (ii) After a period of food deprivation, you feel 'empty' sensations in your stomach. (iii) You can touch your nose with your finger when your eyes are shut.

Question 2.3 (*Objectives 2.6 and 2.7*)
For each of the following statements (a) to (d) decide, first, whether part (i) is consistent with the concept of an FAP and, second, whether or not it remains consistent in the light of the information given in part (ii). Give a brief rationale for your answer.

(a) (i) Males in a given species of grasshopper produce very stereotyped calls. (ii) During development, male grasshoppers have no opportunity to hear their species' call.

(b) (i) Males in a given species of bird produce a song that is species-specific. (ii) Males only produce species-specific song if they have heard their father sing during their development.

(c) (i) Many captive animals show highly stereotyped patterns of behaviour called stereotypies. (ii) Stereotypies are typically highly specific to an individual animal.

(d) (i) Sea slugs show a stereotyped pattern of behaviour when they escape from a predator. (ii) The excised brain of a sea slug generates nerve impulses similar to those observed in an escaping, intact animal.

Question 2.4 (*Objective 2.8*)
In spring and summer songbirds typically sing vigorously in the morning—the 'dawn chorus'. The song of an individual bird could be a response to either (a) the increase in ambient light, or (b) the song of other birds. How would you distinguish between these two hypotheses?

Question 2.5 (*Objectives 2.9 and 2.11*)
Which of the following statements are true and which are false? (a) An egg is a sign stimulus that elicits egg-retrieval in the herring gull. (b) Herring gulls respond more strongly to an object that is large, oval and green compared with one that is large, rectangular and brown because of heterogeneous summation. (c) A supernormal stimulus is the stimulus in the natural environment that is most effective in eliciting a particular response.

Question 2.6 (*Objectives 2.1, 2.13, 2.14 and 2.15*)
Give definitions of the following: (a) releaser; (b) stimulus–response chain; (c) consummatory stimulus; (d) taxis; (e) kinesis; (f) pheromone; (g) hormone.

Question 2.7 (*Objectives 2.16 and 2.19*)
Among frogs, males typically produce calls that are highly species-specific, for example in terms of the sound frequency (pitch) which is most prominent in the call. What would you conclude about the nature of stimulus filtering in the auditory system of females if (a) females in all frog species had very similar ears, (b) the ears of female frogs of different species contained receptors specifically 'tuned' to the frequencies that are most prominent in the calls of males of their own species?

References

Baerends, G. P. and Kruijt, J. P. (1973) Stimulus selection, in R. A. Hinde and J. Stevenson-Hinde (eds) *Constraints on Learning: Limitations and Predispositions*, Academic Press, pp. 23–50.

Dawkins, M. (1971) Perceptual changes in chicks: another look at the 'search image' concept, *Animal Behaviour*, **19**, pp. 566–574.

Ewert, J-P. (1974) The neural basis of visually guided behaviour, *Scientific American*, **230**(3), pp. 34–42.

Ewert, J-P. (1985) Concepts in vertebrate neuroethology, *Animal Behaviour*, **33**, pp. 1–29.

Sevenster-Bol, A. C. A. (1962) On the causation of drive reduction after a consummatory act (in *Gasterosteus aculeatus* L.). *Archives néerlandaises de zoologie*, **15**, pp. 175–236.

Further reading

Dawkins, M. S. (1983) The organization of motor patterns, in T. R. Halliday and P. J. B. Slater (eds) *Animal Behaviour, Vol. 1. Causes and Effects*, Blackwell Scientific Publications, Oxford, pp. 75–99.

Land, M. F. (1983) Sensory stimuli and behaviour, in T. R. Halliday and P. J. B. Slater (eds) *Animal Behaviour, Vol. 1. Causes and Effects*, Blackwell Scientific Publications, Oxford, pp.11–39.

CHAPTER 3
GENES AND BEHAVIOUR

3.1   Introduction

> ...it is not appropriate to speak of characteristics as genetically or environmentally determined: all features are influenced both by the genes and by the environment. (S. A. Barnett (1981) *Modern Ethology: the Science of Animal Behavior*, Oxford University Press)

In the previous chapters the behavioural similarities between animals of the same species (conspecifics) was emphasized. Thus, given appropriate stimuli, weaver birds build weaver bird nests, male sticklebacks perform zigzag movements and male robins attack. A more detailed examination of these patterns of behaviour, however, reveals that one weaver bird's nest can be distinguished from another's, that the form and duration of the zigzags varies from stickleback to stickleback and that different robins attack in different ways. These differences, or variation, in the way conspecifics perform the same behaviour can sometimes be accounted for by short-term changes in the internal state of the animal in question, i.e. its motivation. (This was the suggestion, you remember, that was put forward to account for the differences in spermatophore deposition in newts in Chapter 1.) Often, though, such differences do not result from short-term changes and are relatively stable characteristics of an individual animal. (Keepers of domestic animals, from mice to race-horses, will readily attest to stable differences between individuals, by referring to the 'personalities' of their charges.) Stable differences between individuals are the result of two interacting factors; what an individual inherited from its parents—its genetic material—and the environment in which that individual developed. The influence of the genetic material on behavioural variation is the subject of this chapter, whereas a consideration of the developmental environment on behavioural variation is deferred until Chapter 5.

So, how does the genetic material influence behaviour: what is the relationship between an organism's genetic material and its behaviour? Before we begin to answer that question, though, it is necessary to examine the genetic material in more detail and introduce some basic genetics and genetic terminology. So the next section presents some biology and addresses the questions of what exactly is the genetic material, and how does the genetic material get used?

(If you are familiar with genetics you could move on to Section 3.3.)

3.2 Genetics

3.2.1 From parent to offspring

We all know what people look like. We also know that when people reproduce they produce human babies, not kittens or tortoises. When dogs reproduce they produce puppies, and to be sure that Labrador puppies are produced the Labrador bitch must mate with a Labrador dog. So much is common knowledge. It is also common knowledge that the main part of reproduction is getting a sperm to fertilize an egg. The fertilized egg (the **zygote**) then grows and develops to become the baby, the puppy or the tortoise.

The question to be addressed is, if the only thing adults pass to their offspring is a **gamete** (a sperm from the father and an egg from the mother), how can there be sufficient information in those two gametes to ensure that the offspring looks like its parents (i.e. is a human baby rather than a kitten)? Furthermore, what sort of information is it and how does it work?

These are the central questions of genetics, and are crucial to an understanding of functional and evolutionary explanations of behaviour, not to mention a fair chunk of medicine and psychology! The answers to these questions involve **proteins** and the **genetic code**.

3.2.2 Proteins and the genetic code

Every species of living organism has its own unique and characteristic set of proteins, for example humans contain approximately 100 000 different types of protein. Proteins have many and varied functions, all of which are necessary for life. At a gross level, for instance, muscle consists of a number of proteins. At a microscopic level, proteins contribute to the shape of individual cells and structures (e.g. blood vessels). Proteins act as 'messengers' in the organism, being released from cells in one part and affecting the activity of cells in another part. Some proteins are **enzymes**. Enzymes control the rate at which the myriad chemical reactions in cells take place; most chemical reactions in cells would be very slow or not occur at all without enzymes. In the conversion of phenylalanine to tyrosine say (a reaction you will come across later in this chapter), an enzyme controls the rate at which tyrosine is produced.

Proteins are made of **amino acids** joined one to another into a string. Twenty or so amino acids are commonly found in living organisms and, by arranging different numbers of them in different orders, so different proteins can be made. (It is worth noting that a string of only six amino acids could come in one of 64 million forms! The number of possible proteins is enormous.) However, a protein is not simply a string of amino acids: a protein also has a unique three-dimensional shape. The shape arises because a particular string of amino acids folds up on itself in a particular way. The precise shape of a protein is very often decisive in the functioning of that protein: if the protein is the wrong shape, its functioning will be impaired, or worse, it will not function at all. A mistake in joining up the amino acids, so that just one is in the wrong place or missing, can alter the shape, and result in a protein which simply cannot do its job. It is therefore—literally—vital that proteins are made accurately.

The exact order in which the amino acids are joined together is dictated by the genetic material by means of the genetic code. The genetic material consists of an extremely long molecule called **deoxyribonucleic acid (DNA)**. DNA (Figure 3.1) is formed from two strands joined together. Each strand consists of a 'backbone' made of alternating ribose sugar and phosphate molecules, with a molecule called a **base** attached to each sugar molecule. The two strands are linked together by bonds between opposite bases. There are four types of base and, in any one strand, they can occur in any sequence, although the actual order in the genetic material of a given species is remarkably stable. The exact sequence of bases in the DNA—the genetic code—determines the exact sequence of amino acids in a protein, and as you have seen, the exact sequence of amino acids in a protein determines whether or not that protein can function by affecting its shape and composition. It is the precise sequence of bases in the DNA that is passed from generation to generation in the gametes.

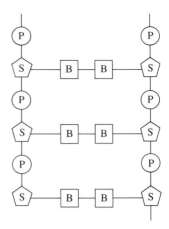

Figure 3.1   Diagram of the molecular structure of DNA. P: phosphate molecule, S: ribose sugar molecule, B: base molecule.

3.2.3   Genes and chromosomes

Virtually all the cells of an organism contain a complete, and exact, copy of the genetic material found in the zygote (a necessary requirement because each cell is producing and breaking down its own proteins continuously). However, not all of the DNA in a cell codes for protein; some DNA seems to do nothing at all, and other bits of DNA control the decoding process, like where to begin and end decoding.

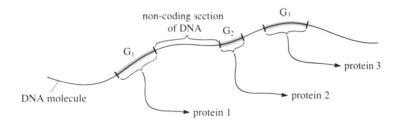

Figure 3.2   A section of DNA showing three genes (G1 to G3), each of which codes for one of the three proteins (protein 1 to protein 3). Note that the genes are of unequal length and that they are separated by lengths of DNA that do not code for protein.

A section of DNA that codes for a protein is called a **gene** (Figure 3.2). (Not all proteins are coded for by a single section of DNA, but this is a complication that you do not need to concern yourself with here.) All the genes contained within the cells of an organism are called the **genotype** of that individual. As there are many thousands of proteins, and each protein is coded for by a particular gene, it follows that there must be many thousands of genes. These genes are arranged so that there are very many genes in one molecule of DNA, strung together in a precise order and interspersed with the non-coding sections of DNA (Figure 3.2). Each different DNA molecule comprises a set of different genes.

When a molecule of DNA is not being decoded, its two strands wind around each other to form a double helix shape (Figure 3.3), with the double helix itself being wrapped in protein. This larger structure, of double helix plus protein, is called a **chromosome**.

Figure 3.3   The double helical shape of DNA.

The number of chromosomes is the same in virtually every cell of an organism, and in every normal individual of the same species, but it varies from species to species (Table 3.1). Furthermore, chromosomes (other than the sex chromosomes—see Section 3.2.4) occur in matching—or homologous—pairs, so a cell containing 46 chromosomes really only has 23 kinds of different chromosome (Table 3.1).

Table 3.1 The number of chromosomes per cell and the number of pairs of chromosomes per cell.

	Number of chromosomes per cell	Pairs of chromosomes per cell
Fruit-fly (*Drosophila*)	8	4
Human (*Homo sapiens*)	46	23
Common frog (*Rana temporaria*)	26	13
Honey-bee (*Apis mellifera*)	32	16
Domestic chicken (*Gallus gallus domesticus*)	78	39

Any given chromosome will always consist of particular genes and those particular genes will each occupy their own particular place on the chromosome (i.e. the order of genes along a chromosome is important). Thus, each gene has its place. Now, because a gene at a particular place always produces the same protein, it always affects the same **character** (any aspect of an individual's behaviour, physiology or morphology), or set of characters, e.g. eye colour. But that gene can come in different forms, so that although each different form still affects eye colour, it can result in a *different* eye colour, e.g. blue eyes or brown eyes. These different forms of a gene are called **alleles**. There are more than 160 known alleles of the gene that codes for the blood protein haemoglobin, for example. Needless to say, places on a chromosome are not called anything as simple as places, they are called loci (singular = **locus**). In a way, a chromosome can be thought of as a series of loci, and at each locus there is an allele of a particular gene. So 'each gene has its place' can be re-stated as 'each allele has its locus'.

Figure 3.4 illustrates a matching pair of chromosomes. Four loci are shown, numbered 11, 12, 13 and 14. At locus 11, gene Q is in form Q on one chromosome and in form q on the other. (By convention genes are referred to by an abbreviation, often the initial letter of the character affected by the gene, and printed in italics.) At locus 13, gene D is in form D on both chromosomes. In the former situation, where different alleles are on each chromosome, the individual is said to be **heterozygous** at that locus; in the latter situation, where the alleles are the same on each chromosome, the individual is said to be **homozygous** at that locus.

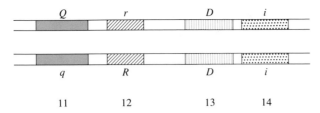

Figure 3.4   Diagram of a section of two homologous chromosomes, showing four gene loci (11 to 14). Loci 11 and 12 are heterozygous; there are different alleles at the same locus on the two chromosomes. Loci 13 and 14 are homozygous; both chromosomes carry the same allele.

☐   Would an individual that is heterozygous at every locus carry all the alleles found in its species?

■   No. Heterozygotes carry two different alleles at each locus but a gene may have more than two alleles—remember the 160 alleles of the gene that codes for haemoglobin. For any one gene, a maximum of only two alleles can be present within a cell because there are only two appropriate loci, one on each chromosome of a matching pair.

In the homozygous condition, the question of how an animal's **phenotype** (all the aspects of its behaviour, morphology and physiology, i.e. the sum of all its characters) will turn out does not present any problems. The two alleles are the same and if that allele is for blue eyes, say, then the animal will have blue eyes. Things are not quite so straightforward in the heterozygous condition though. What happens to an animal's phenotype in the heterozygous condition depends on the particular alleles involved. For many kinds of gene the animal's phenotype is intermediate between that of animals homozygous for one allele and those homozygous for the other allele, but for many other kinds of gene this does not happen. Instead the animal shows the phenotypic character associated with one of the alleles and shows no sign of the phenotypic character associated with the other allele. For example, there are several alleles affecting human eye colour. People with two 'brown' alleles develop brown eyes while people with two 'blue' alleles develop blue eyes. But those with one 'brown' allele and one 'blue' allele develop brown eyes—there is no sign of the 'blue' allele in the phenotype and it is not possible to tell, simply by looking at them, that they have a 'blue' allele. People who have a 'brown' allele, even if they only have one, are able to make a protein necessary to make a particular brown pigment and, as a result, they develop brown eyes. People with two 'blue' alleles are unable to make the brown pigment at all and in its absence their eyes look blue. In instances like this the allele whose effect is apparent is said to be **dominant** to the other allele while the allele whose effect is hidden is said to be **recessive**. Dominance and recessiveness are not absolute properties of alleles—allele A may be dominant to allele B but recessive to allele C, for example.

3.2.4   Gametes

In each cell, each matching pair of chromosomes consists of genes different from any other matching pair of chromosomes. It is also the case that each matching pair of chromosomes has a distinctive size and shape (Figure 3.5).

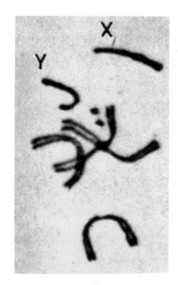

Figure 3.5   Photograph of the chromosomes of *Drosophila melanogaster*, which has four chromosome pairs. Note that each chromosome has a distinctive shape, similar to that of its partner. X and Y are the (non-matching) sex chromosomes.

One chromosome of a matching pair is inherited from the mother—the maternal chromosome—and the other chromosome of a matching pair is inherited from the father—the paternal chromosome. (Sex chromosomes are the exception to the rule. Generally, females have a matching pair of X chromosomes, but males have a non-matching pair of one X and one Y chromosome, or, in some cases, a Y chromosome only. In some organisms, such as birds, this is the other way around, females being XY and males XX. This distinction is important but not essential to the present discussion, which ignores the special problems of the sex chromosomes.)

Somehow, then, each homologous pair of chromosomes must be reduced to a single chromosome in each gamete (else it would not be possible to inherit just one chromosome of each homologous pair from each parent). An organism that has four matching pairs of chromosomes in each cell, for example, would have four single chromosomes in each gamete, one from each pair. Which particular chromosome from a matching pair goes into each gamete is totally random, and each pair of chromosomes is divided up independently of all the others. So, an organism with 4 pairs of chromosomes could produce 16 possible gametes (Figure 3.6). This feature of gamete formation is called **independent** (or **random**) **assortment**. As a result of independent assortment, each individual inherits a random mixture of the chromosomes that originally came from its maternal grandfather and the chromosomes that originally came from its maternal grandmother. Similarly, each individual inherits a random mixture of the chromosomes that originally came from its paternal grandfather and grandmother.

In fact, many more gamete types can be produced by each individual than suggested above because of another process called **crossing over** (Figure 3.7). Crossing over, which together with independent assortment is known as **recombination**, results in the parent's alleles being 'mixed up' before being passed on to the offspring. The process of gamete formation (**meiosis**) consists of two stages. In the first stage the two chromosomes that constitute a matching pair become aligned and then a section of DNA from one chromosome crosses with the matching section of DNA from the other chromosome. At each crossing point, the chromosomes exchange equivalent sections of DNA, thus exchanging some of their alleles and forming chromosomes of the same size and shape as before but with new allele combinations. There may be more than one crossing point for each chromosome pair and the points at which the exchanges happen are different each time meiosis takes place. So the chances of two gametes having the same combination of alleles are extremely small, and the number of possible gamete types is almost infinite.

It is in the second stage of meiosis that pairs of chromosomes separate, with each chromosome of a pair ending up in one of two different gametes. This separation results in each gamete containing one chromosome of each type (Figure 3.6).

Meiosis thus results in each gamete being unique, containing one chromosome of each type, and with each chromosome being a novel combination of alleles.

Thus each gamete has half the number of chromosomes contained by ordinary cells. When one gamete, a sperm, fertilizes another, an egg (or ovum), a full set of chromosomes is restored (Figure 3.8).

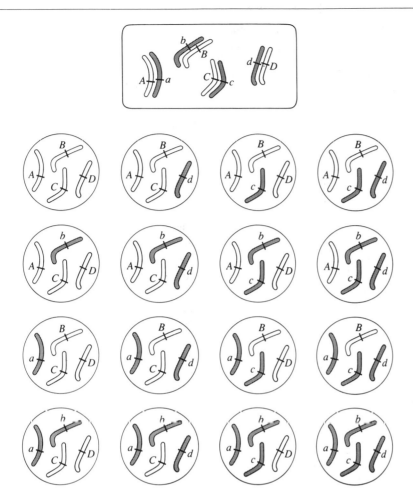

Figure 3.6 Diagram showing how independent assortment of chromosomes can produce a large number of different gene combinations in gametes. In the parent, there are four pairs of chromosomes, each heterozygous at a particular locus. Maternal (white) and paternal (shaded) chromosomes assort independently into gametes, yielding 16 possible unique combinations of the 8 chromosomes.

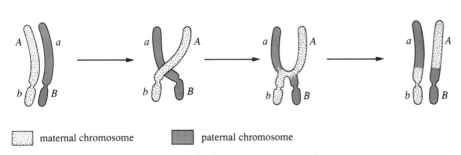

maternal chromosome   paternal chromosome

Figure 3.7 Diagram showing how crossing over between homologous chromosomes can lead to the reassortment of alleles. The paternal and maternal members of a homologous pair first come together side by side. One chromosome then comes to lie across the other; the chromosomes break and rejoin so that their homologous portions are exchanged. When they separate, genetic material has been exchanged so that two new chromosomes are formed, carrying different combinations of alleles from the original pair.

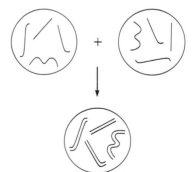

Figure 3.8 Fusion of two gametes, each containing half the parental number of chromosomes, produces a zygote in which the parental chromosome number is restored.

The zygote begins to develop by dividing in two. This process is called **mitosis**. The two resulting daughter cells also divide in two, as do their daughters in turn and so on. Each daughter cell contains a full complement of DNA and no novel gene combinations (recombination does not occur in mitosis). Although each cell has the potential to produce all the proteins characteristic of the organism, only some of the possible proteins that could be made by the cell are made at any particular time. As development in multicellular organisms proceeds, and cells differentiate to become muscle cells or nerve cells or any other kind of cell, so different proteins are made while others stop being made until the final form of each cell is reached. At this point, cells that are different produce some similar and some dissimilar proteins.

3.2.5 Mutations

Each gene codes for a protein (Section 3.2.3). If something should happen to the gene so that the sequence of bases is altered in some way then the protein too would be altered. Even very small changes in the sequence of bases can have profound effects on the protein formed. A change to a gene, called a **mutation**, can result from, for example, a mistake in meiosis or mitosis, or from an alteration to the DNA caused by factors such as chemicals or radiation. If a mutation occurs in a single cell in the adult body, there may be no obvious effects. But if the mutation occurs in the formation of the gametes, and one of those mutant gametes is involved in zygote formation, then all the cells that develop from the zygote, i.e. all the cells of the embryo, the fetus and ultimately the adult, will contain the mutation. This might or might not be serious, depending on the type of mutation. Mutations can be beneficial, neutral or deleterious.

A small change to a gene might result in a change in only one amino acid in the whole sequence of amino acids that make up the protein. Alternatively, a mutation might result in many of the amino acids being changed. The ultimate outcome of the mutation depends on the type of change effected in the sequence of amino acids.

A **neutral mutation** occurs when any change in the protein resulting from the change in the gene does not measurably affect the functions of the protein. It is sometimes possible to substitute one amino acid for another and not alter the functioning of the protein. These neutral mutations do not measurably affect the individual's phenotype (they do, of course, affect the phenotype of the protein itself).

A deleterious mutation adversely affects the individual carrying it in some way. The effects can range from very minor ones to those that are lethal, i.e. that result in the organism dying at some stage in its development before it reaches sexual maturity. A mutation may be lethal if the protein affected is one essential to life and the mutant protein is unable to function properly or at all.

☐   How might a mutation be beneficial?

■   Changes to the protein that the mutated gene codes for might make it more efficient, i.e. by using less energy or acting more quickly, or the mutation might make the protein stronger. Alternatively a novel protein might be produced that could, for example, break down a previously toxic substance.

Because proteins are such complex molecules and their structure and function are so closely integrated, however, any change in the amino acids that make up a protein is probably more likely to be deleterious than beneficial.

Whether a mutation is beneficial or deleterious, it results in an altered protein being produced and hence an altered phenotype. Individuals with an altered phenotype are sometimes referred to as **mutants**.

You are now equipped with an armoury of genetic jargon. Refer back to it (or to the glossary) when the need arises as you work your way through the examples that follow.

The central question with which the rest of this chapter is concerned is: 'What is the relationship between the genetic material and behaviour?'. Two examples in particular, phenylketonurea and yellow *Drosophila*, recur in the following sections as the relationship between genetics and behaviour is expounded.

Summary of Section 3.2

Proteins form the structure and contribute to the function of cells and, ultimately, organisms. The exact sequence of amino acids that comprises a protein is dictated by a sequence of bases in the DNA called a gene. Genetic variation arises through the shuffling, by recombination, of genes at meiosis and through mutation.

3.3   Effects of single genes

In this section the simplest relationship between genes and behaviour is explored, namely where a change in one gene affects one aspect of behaviour.

3.3.1   Hygienic and non-hygienic honey-bees

Honey-bee (*Apis mellifera*) eggs are laid at a prodigious rate by a single reproductive female in each colony, the queen. Each egg is laid in its own hexagonal cell within the comb. The egg hatches into a larva and is provided with a food supply of nectar by sterile females, the workers. By about its fifth day of life, the larva is sealed up in its cell where it normally turns into a pupa, which in turn becomes an adult bee and emerges from the cell. Occasionally, however, things go wrong; the larva can be attacked and killed by a bacterial infection which can spread through the wax of the combs and produce a severe epidemic. It turns out that some colonies cope better with an infection than others and this, in turn, appears to be due to a behavioural difference.

In some colonies, known as 'hygienic' colonies, a worker that comes across a cell containing a dead larva or pupa will uncap the cell in which the dead larva or pupa is lying and remove the corpse. Workers from other 'non-hygienic' colonies do not uncap the cell or remove the corpse. The biologist Walter Rothenbuhler, working in Ohio, studied the genetic basis of the difference between these two types of colony (Rothenbuhler, 1964).

Rothenbuhler used a procedure known as inbreeding, whereby closely related animals are mated together, e.g. brother with sister, or cousin with cousin, for several generations. This technique produces animals of very similar genotype collectively known as a **strain**. Animals from a particular strain always breed true, i.e. they are homozygous. Rothenbuhler thus developed a 'hygienic' strain which only produced hygienic colonies and a 'non-hygienic' strain that only produced non-hygienic colonies. He then took queen bees and inseminated them artificially each with the sperm from a single drone (a male bee). When the queen and the drone were from different strains, all the offspring colonies were found to be non-hygienic. This result can be explained by assuming that only one gene is involved.

☐ Given that all the offspring colonies from a hygienic × non-hygienic mating were non-hygienic, does this result mean that the non-hygienic allele was dominant or recessive to the hygienic allele?

■ Dominant, because even though all the offspring must have had an allele of each type (i.e. they were heterozygous), their phenotype was non-hygienic.

The next stage in his genetic analysis was to **back-cross** offspring from this first mating with individuals from the parental hygienic strain. The result was slightly surprising and is shown in Figure 3.9.

Out of the 29 back-cross colonies he produced, there were eight non-hygienic and six hygienic colonies. In addition there were two new types of colony: nine colonies uncapped the infected cell but did not remove the infected larva, and six colonies did not uncap the infected cells but would remove the infected larvae if

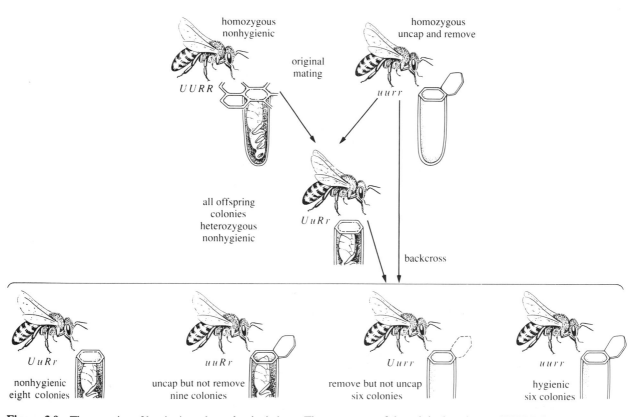

Figure 3.9 The genetics of hygienic and non-hygienic bees. The genotypes of the original strains are $UURR$ for non-hygienic, $uurr$ for the hygienic strain.

the cells were first uncapped. These results showed that different genes must control the uncapping and removal behaviours, which Rothenbuhler called, unsurprisingly, *U* (uncap) and *R* (remove) respectively. Both removal and uncapping are 'hygienic' behaviours and so the alleles for these must be recessive (depicted as lower case *u* and *r*) to explain the results of the first mating. (If the alleles for hygienic behaviours were dominant, then all the bees resulting from the first mating would show these hygienic behaviours.) Thus the recessive allele for each of these behaviours must be on both chromosomes of a matching pair in the hygienic strain, i.e. the genotype must be *u u r r* in the hygienic strain.

☐   What is the genotype called where both alleles on corresponding chromosomes are the same?

■   Homozygous.

☐   What is the converse condition, where the alleles on matching chromosomes are different?

■   Heterozygous.

In this example two behaviour patterns are seen to be associated with two loci. When both loci are homozygous for the recessive alleles, beneficial hygienic behaviour is found. If either locus is heterozygous then some form of non-hygienic behaviour is found.

3.3.2   Mating in yellow Drosophila

The fruit-fly (*Drosophila melanogaster*) is a favourite organism of behavioural geneticists. It breeds quickly, it only has four pairs of chromosomes and there are hundreds of known mutations. One such mutation, studied by Margaret Bastock (1956), is called 'yellow body', where the body has alternating yellow and dark brown bands, in contrast to the grey/black bands of normal (or **wild-type** flies).

What Bastock did was to inbreed yellow flies for seven generations to ensure that they were genetically similar. She also inbred wild-type flies for seven generations. She then put a virgin yellow male and a virgin wild-type female aged about four or five days together in controlled conditions for 1 hour—even at four days of age these flies are sexually mature. She also put virgin wild-type males with virgin wild-type females. The results are presented in Figure 3.10 (*overleaf*). Running from left to right is the time from the initial pairing of the male with the female, up to the end of the test at 60 minutes. On the vertical axis is the cumulative mating success, that is the percentage of pairs that have copulated. So, for instance, at 30 minutes only 38% of pairs with yellow males have copulated compared with 76% of pairs with wild-type males.

☐   What can you conclude about the mating of yellow males compared with wild-type males?

■   Yellow males are slower to achieve mating than wild-type males. Nearly 30% of wild-type males had mated in five minutes, but only 4% of yellow males had.

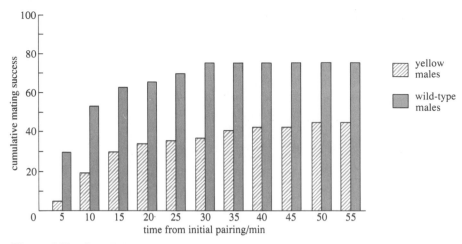

Figure 3.10 Cumulative percentage of successful matings which occur at different time intervals by wild-type and yellow males when paired with wild-type females.

The main conclusion is that the yellow allele increases the time yellow males take to mate compared with wild-type males. In this example then the yellow males can still perform the necessary behaviours—after all some 46% do mate—but they are slower to achieve mating than the wild-type males. This could be for a number of reasons, e.g. their actual mating behaviour could be slowed down or they could be less attractive to females than wild-type males. Thus, in contrast to the uncapping and removal alleles which affect whether a behaviour will occur or not, the yellow allele affects the speed with which mating is achieved rather than the actual performance of the behaviour.

3.3.3 Phenylketonuria

Phenylalanine is an amino acid which is abundant in the diet. In most people, as the first step of the metabolic breakdown of phenylalanine, it is converted into tyrosine. However, some newborn babies are unable to convert phenylalanine into tyrosine, which results in the accumulation of high levels of phenylalanine in the blood, and subsequently in the urine: a condition known as **phenylketonuria (PKU)**. High levels of phenylalanine disturb normal development in such a way that infants who suffer from PKU have behavioural disorders and severe mental retardation.

The metabolic and genetic origins of this disease were discovered in 1934 by the Norwegian physician and biochemist A. Følling. He showed that PKU patients fail to make an enzyme, phenylalanine hydroxylase, because of a genetic deficiency—a mutation. (They make an inactive version of the enzyme.) This is a natural mutation, which means that it arose spontaneously in the population, not as the result of artificial manipulation.

Enzymes, remember, are proteins. So in this example, a mutation results in a defective protein. The defective protein is unable to convert phenylalanine to tyrosine and so excessive amounts of phenylalanine remain in the blood. Just to complete the picture, the mutant gene is recessive to the normal gene, so PKU patients must all be homozygous for the mutant gene. (Only under homozygous conditions is a recessive gene expressed.)

In this example a single mutation results in an altered protein, which results in altered behaviour.

3.4   Pleiotropy

The discussion so far has implied that a gene has one effect and only one effect. In the sense that one gene codes for one protein this is clearly the case but at the level of the phenotype the situation is more complicated. Consider the yellow allele discussed in Section 3.3.2

☐   What two phenotypic differences do you know of between males with the yellow allele and males with a wild-type allele?

■   One difference is that yellow males are slower at mating than wild-type males. The other difference is that yellow males have yellow stripes on their bodies.

The yellow allele, then, has more than one effect on the phenotype—an example of what geneticists call **pleiotropy**. Pleiotropy is the norm rather than the exception. The cellular mechanisms controlled by enzymes, the metabolic pathways, are so complex and inter-related that a change in one enzyme is likely to affect several pathways and thus several phenotypic characters.

The inter-relatedness of metabolic pathways is well illustrated by phenylketonuria. This mutation affects the enzyme phenylalanine hydroxylase with the result that phenylalanine is not converted into tyrosine. The next stage in this pathway is the conversion of tyrosine into 3,4-dihydroxyphenylalanine, a chemical better known as DOPA. DOPA itself can be converted into both dopamine, an important chemical messenger in the nervous system which will be discussed in Book 2, and to melanin, an important pigment (see Figure 3.11).

It follows from this that disturbances to both nervous system functioning and colour might be expected in people with PKU. Behavioural disorders were mentioned as a consequence of PKU in Section 3.3.3; pale skin and fair hair are other common symptoms.

In PKU then, one very specific mutation to one gene affects several phenotypic characters, though no one knows the exact nature of all the links between the gene and the disorder.

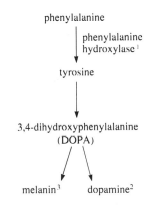

Figure 3.11   Pathways showing the biochemical breakdown of phenylalanine.
Notes
1   This enzyme is defective in phenylketonuria.
2   Dopamine is an important chemical in the nervous system.
3   Melanin is an important pigment of skin and hair.

3.5   Polygenic effects

In the preceding examples a clear relationship was shown between the presence of a particular allele in an organism and a particular characteristic of the phenotype. Such a clear relationship is uncommon as many phenotypic characters are not associated with a single gene, but rather are the result of many genes interacting. That is, the character in question is **polygenic**.

The examples that follow illustrate the techniques used to investigate polygenic characters. The main purpose of these investigations is to show that it is possible to deduce genetic influences on certain characters, and thus that variation between individuals in a particular character is due to variation in the genotypes of those individuals.

The three principal techniques used to establish that differences in genotype are associated with differences in behaviour are the use of behavioural variants, selective breeding and strain differences.

3.5.1 Behavioural variants

Closely related species sometimes have conspicuously different behaviour patterns. If these species are capable of interbreeding then it is possible to study genetic influences on the patterns of behaviour.

One such example was studied by William Dilger (1962) at Cornell University. He examined the nest building behaviour of two species of lovebird, the peach faced lovebird (*Agapornis roseicollis*), and Fischer's lovebird (*Agapornis fischeri*). Birds of both species carry strips of nesting material back to the nest, but whereas the peach-faced lovebird carefully tucks the strips into the feathers of its rump, Fischer's lovebird carries the nesting material in its beak.

The offspring of a mating between these two species, a **hybrid**, has great difficulty carrying any nesting material at all. It tears off strips of nesting material, and then tucks the material into its feathers. However, it does so very unsuccessfully, either tucking the material into the wrong feathers, or not securing the material, or failing to release the material from its beak when it is tucked in. Eventually the hybrids can learn to cope and carry nesting material but only after several years.

The hybrids then, have not inherited the full behaviour pattern of either parent, but rather have inherited components of the behaviour pattern involved in carrying nesting material of one parent and components of the behaviour pattern of the other parent.

The conclusion from this study is that the differences in the behaviour patterns involved in carrying nest material of the two original species are associated with differences in their genotype.

In a similar way it is possible to tackle some of the genetic parameters underlying song in crickets.

Male crickets use a calling song to attract females over long distances. The sound is produced by the rhythmic opening and closing of specialized forewings. A sound pulse is produced with each closing stroke of the wings, while the opening stroke is silent. The calling songs are remarkably stereotyped among members of a local population, but they are different between species. The differences occur primarily in the intervals between pulses and the duration of the pulses.

The songs of two species of cricket, *Teleogryllus oceanicus* and *Teleogryllus commodus,* are shown in Figure 3.12 as **sonagrams**. A sonagram presents the sound made by an animal as a graph, with horizontal and vertical components. Those in Figure 3.12 show sound intensity (loudness) on the vertical axis plotted against time on the horizontal axis. The song of these crickets is a trill consisting of a rapid series of discrete 'chirps'.

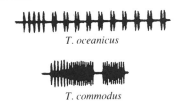

T. oceanicus

T. commodus

T. commodus female × *T. oceanus* male   *T. oceanus* female × *T. commodus* male

Figure 3.12 Pictorial representations (sonagrams) of the songs of male crickets. *Top:* the songs of two species, *Teleogryllus oceanicus* and *Teleogryllus commodus*. *Below:* the calls of two males that are hybrids between these two species: *left:* the offspring of a *T. commodus* female and a *T. oceanicus* male; *right:* the offspring of a *T. oceanicus* female and a *T. commodus* male. A sonagram shows sound intensity (loudness) on the vertical axis, against time on the horizontal axis; the time-scale in each of the sonagrams is the same.

The song of the male is shown; females do not sing in these species. There are two possible hybrids between these two species of cricket, depending on which species was the male parent and which the female.

☐   Look at Figure 3.12. Are the songs of the hybrids the same as or different from that of (a) their male parent, (b) each other?

■   The songs of these two hybrids differ and are also distinctly different from either parental song.

Clearly then the genotypes of the parents influence the phenotypic character, song. Furthermore, as song is the product of friction mechanisms, wing movement, muscles and nerves it follows that song must be polygenic, influenced by many genes.

Males produce species typical song which would be of no use in attracting a mate unless females were preferentially attracted to the song of their own species. It turns out that females do indeed prefer the song of their own species to that of another species. The intriguing question now is, what song do hybrid females prefer: the song of the species to which their mother belongs, the song of the species to which their father belongs or some other song?

Ronald Hoy and Robert Paul employed an ingenious piece of apparatus to answer this question (Figure 3.13).

The cricket holds on to the loop and walks along it, thereby turning the loop around. Song of different crickets is played to her left and right sides and when she tries to turn to approach the sound she twists the loop and continues walking. What Hoy and Paul found was that female hybrids derived from crossing *T. commodus* males with *T. oceanicus* females preferred the song of the male of the same parentage, i.e. *T. commodus* male × *T. oceanicus* female. Female hybrids derived from *T. oceanicus* males and *T. commodus* females preferred the song of the male of that parentage, i.e. *T. oceanicus* male × *T. commodus* female.

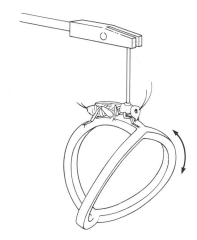

Figure 3.13 Apparatus used to detect the direction of movement of a female cricket. The female is held from above and she holds the Y-shaped structure with her feet. The female is fixed in position between two speakers and any movement she attempts to make towards one or other speaker causes movement by the Y-structure, not by the female herself.

It would seem that variation in the attractiveness of the song to the female is under the same genetic influence as the production of song in the male.

3.5.2 Selective breeding

Studies of behavioural variants can obviously only work if there are behavioural variants to study in the first place. So another technique is to try to separate behavioural variants from within a population. Selective breeding does this by relying on the natural variation within a species and then only allowing certain selected individuals to breed, repeating the selection process in each generation.

Robert Tryon, working in California in the 1930s, undertook a selective breeding experiment which began by classifying rats as either 'maze-bright' or 'maze-dull'. In this experiment, Tryon put hungry rats in the dark in a maze which had a path from a start box to a goal box containing food. The maze also had lots of blind alleys. Each rat was placed in the maze nineteen times and the number of times it entered a blind alley before it found the food (the number of 'errors') was counted each time. Rats that were good at the task (maze-bright) would learn not to run up blind alleys on their way to the goal box and so had a low error score. Rats that were poor at the task (maze-dull) would run up lots of blind alleys and had a high error score. Initially, Tryon tested 142 rats of both sexes which he called the parent generation. He then allowed the rats with the fewest errors to mate together and the rats with the most errors to mate together, to produce two first generations, a low-error strain and a high-error strain. The first generation rats were tested in the maze in the same way as the parent generation, and those rats with the fewest errors in the low-error strain were mated together and the rats with the most errors in the high-error strain were mated together to produce a second generation. He repeated this for seven generations each time ensuring that, in the low-error strain, low-error rats only mated with low-error rats and, in the high-error strain, that high-error rats mated only with high-error rats. The results for the parent generation and the offspring of the seventh generation rats are shown in Figure 3.14 (Tryon, 1940).

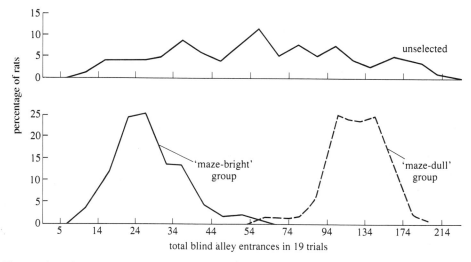

Figure 3.14 The number of errors made in a maze by three strains of rats: unselected controls, maze-bright and maze-dull. The percentage of rats making a given number of errors is shown on the vertical axis, while the number of errors is shown along the horizontal axis. There were 142 rats in the unselected sample, 90 in the maze-bright group and 72 in the maze-dull group.

Quite clearly, Tryon had two distinct strains of rats, those with an overall low error score and those with an overall high error score, and these strains were the product not of differential training, but of selective breeding. The difference between the two strains was specific to the maze used by Tryon; their performance in other designs of maze or in other tests of learning was not different. Thus Tryon had not selected for 'general intelligence', but for ability to run a particular maze.

☐ The maze-bright and maze-dull strains differ in phenotype. Do they also differ in genotype?

■ Yes.

☐ How did the difference in genotype come about?

■ In the parent population, the small differences between individual rats in maze running must have been due, at last in part, to small differences in genotype. Selective breeding accentuated the small differences in genotype, making them slightly bigger differences in genotype and thus increasing the difference in maze running between the two strains.

Behavioural differences that respond to selective breeding must be due, at least in part, to differences in genotype.

3.5.3 Strain differences

In the previous example two strains of rat were created by breeding together rats that had the same behavioural character. Strains can also be created simply by inbreeding from a small parent population (i.e. without selecting for any particular character). Such inbred strains are composed of animals that have very similar genotypes, but have genotypes that are different from the animals of other strains. If one inbred strain is subsequently found to differ from another inbred strain of the same species in some aspect of behaviour even though they were raised under identical environmental conditions, then their different genotype must have contributed to the behavioural difference. Behavioural differences between inbred strains have been studied using mice.

In a comparison of four strains of mice, all raised under the same conditions in their laboratory in New Jersey, Martin Hahn and Sonja Haber measured a number of aspects of aggressive behaviour (e.g. tail lashing, chasing, fighting). The tests involved putting two male mice of the same strain together in an observation cage and recording their aggressive behaviour until either one animal lay on its back, squeaking, or 30 minutes had elapsed. This was repeated with a number of pairs of males from each strain. Scores of an average pair from each strain on one measure of aggressive behaviour, which gives a numerical index of fighting intensity, are presented in Table 3.2.

☐ How does this example support the idea that some differences in behaviour are due to differences in genotype?

■ Different strains have different genotypes. These strains also differ in behaviour. It follows that the difference in behaviour is probably due to the different genotypes.

Table 3.2 Mean fighting intensity in four strains of mice.

Strain	Mean fighting intensity
C	0.6
D	1.8
B	2.5
S	3.0

The evidence presented in this section from studies of behavioural variants, selective breeding and strain differences, supports the view that some differences in behaviour are due to differences in genotype. In the next section we explore how the genotype influences behaviour.

3.6   From genotype to phenotype

In this section the apparently neat relationship between genotype and behavioural phenotype undergoes a series of qualifications. These qualifications are necessary because the relationship between genotype and behavioural phenotype is not neat and it is important to understand some of the reasons why. The absence of a clear, direct relationship requires that the basic premise, i.e. that the genotype influences behaviour, be couched in deliberately cautious terms.

3.6.1   How does the genotype influence behaviour?

Section 3.2 introduced the biology of the gene and the fact that genes produce proteins. Proteins are not behaviour. Proteins may be structures or they may be enzymes or perhaps chemical messengers, but they are not behaviour patterns. So one question that might have been niggling you up to this point is: if genes only produce proteins (and they do), how do genes influence a particular pattern of behaviour? That genes might affect behaviour in a general way is less difficult to understand. Muscle tissue is very largely composed of protein. If there was a mutation in a gene that codes for a muscle protein, say, then all muscle protein in an organism would be affected. The altered muscle protein might not be able to contract smoothly, generating shakes or tremors. All behaviour of an animal with this particular mutation would be subject to shakes and tremors.

So by what means are genes able to affect particular patterns of behaviour? One way of approaching this question is to find out where the genes act. It is possible to investigate the site of action of genes by using **genetic mosaics**. In a normal (that is, non-mosaic) individual that possesses a mutant allele, the allele is found in every cell of the body (with the exception of some of the gametes). In mosaics, however, the mutant allele is present in only some, not all, of the body cells. It is distributed in one or more patches, so that the body consists of a mosaic of normal and mutant tissue. Thus, if the mutant allele changes the colour of the body to yellow, the body surface of the mosaic individual will bear patches of yellow, consisting of cells containing the mutant allele, and patches of normal-coloured tissue, containing the normal allele (Figure 3.15).

Mosaic fruit-flies can be used to investigate the site of gene action in the body in the following way. If it is known that a particular behavioural abnormality is associated with a particular mutant allele, then the technique is to look at a large number of mosaic flies that possess the mutant allele, and discover which part of the body has to contain the mutant allele for the behavioural abnormality to appear. For example, if the mutant allele is thought to affect behaviour by acting on the brain, then flies that have a mutant brain but a normal body are examined. If the mutant allele does indeed have its effect by acting on the brain, these mosaics should show the abnormal behaviour, whereas other mosaics in which the brain is normal but other parts of the body are mutant, should not have the abnormal behaviour.

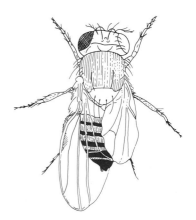

Figure 3.15   An example of a mosaic fruit-fly. The animal's left side is wild-type, while the right side carries genes for white eye and miniature wing.

CHAPTER 3 GENES AND BEHAVIOUR

Seymour Benzer in the 1960s and 1970s used mosaic flies to investigate many different mutant forms. One mutant that Benzer studied was the hyperkinetic mutant, which reacts to ether anaesthesia, not by lying still, which is the response of normal flies, but, at least initially, by shaking all of its legs. Mosaic flies with normal legs but a mutant body were hyperkinetic. However, mosaic flies with mutant legs but otherwise normal bodies, lay still when anaesthesized by ether (Benzer, 1973).

☐ What does this tell you about the site of action of the mutant allele?

■ The mutant allele does not act directly on the legs, if it did, it would be the mosaic flies with mutant legs that were hyperkinetic.

Further research showed that the allele acts on six separate sites, one for each leg, close to each leg, but not part of it, probably situated in the groups of nerves which control each leg.

Another way of finding out how a gene influences behaviour is to establish what the gene does at the biochemical level. A mutant allele on the X chromosome of the mouse is associated with a syndrome called testicular feminization (Tfm). Males with this trait do not generally attack other mice nor elicit attack from them; they are themselves less aggressive than normal males and are perceived as less of a threat than normal males by other mice. The sensory system of these mice is intact and their movements are normal, as is their output of the hormone testosterone. (Testosterone affects the aggressiveness of mice.) It turns out that the gene in question codes for the protein that detects testosterone. The mutant allele associated with Tfm alters the protein so that it is a less effective detector of testosterone. The tissues of the mouse are unable to detect testosterone and so do not respond to it. This means that numerous biochemical pathways that require testosterone as a trigger never become operational. As a result, for example, they never produce the smell typical of the male mouse.

The more complex a behaviour is the more possibilities there are for different genes to exert their influence. For instance aggressive behaviour between two individuals involves not merely the movement of muscles in coordinated sequences, but is also dependent on the appearance (e.g. size, colour and plumage or fur condition) and perhaps smell of those individuals. Thus a gene which influences any one of these characteristics, even indirectly, as in the case of the Tfm mutation, could affect aggressive behaviour.

3.6.2 Individual variation in genetic effects

In the examples of pleiotropy and polygeny, it was shown that differences in genotype were associated with differences in behaviour. Thus flies with the yellow allele were slower maters than flies with the normal allele at the same locus. The differences in mating speed though were not absolute, they were differences in degree; *most* flies with the yellow gene were slower maters than *most* flies with the normal gene. The results of the speed of mating test for yellow *Drosophila* were presented in Figure 3.10.

☐ Ignoring the 20% or so of wild-type males who did not mate, what is the fastest mating speed and the slowest mating speed shown by wild-type flies, to the nearest five minutes?

■ The first wild-type males mated within five minutes and the last mated within 25–30 minutes. The range is therefore of the order of something less than five minutes to up to 30 minutes.

☐ What is the range of mating speeds shown by yellow males?

■ Yellow males mated in something less than five minutes to up to 50 minutes.

Thus although most yellow males are slower than most wild-type males, some yellow males are as fast as the fastest wild-type males. Furthermore, it would not be possible to use speed of mating to distinguish yellow from wild-type males; an individual male that mated in ten minutes could be either wild-type or yellow. The range in behaviour of yellow males means that the presence of the yellow allele in the genotype does not *necessarily* render a fly slow to mate. It means that yellow males are more likely to be slow to mate. Put another way, the yellow allele increases the probability that a fly will be slow to mate.

The probabilistic nature of genetic influences is also found in PKU. One of the symptoms of PKU is elevated levels of phenylalanine in the blood, which can have a debilitating effect if left untreated. However some people can be perfectly normal except for high levels of phenylalanine; in other words, some phenotypes can cope with elevated phenylalanine levels. This raises another important caveat: genes do not act alone. Although individual genes produce individual proteins, those proteins interact to produce the phenotype. The exact effect of any individual protein depends on which other proteins are also present. In the current example, the mutant allele that would otherwise lead to the symptoms of PKU is presumably present with other alleles, the combined effect of which is that the individual has no other symptoms.

Genes, through a series of steps, give rise to proteins which, in turn, affect the biochemical pathways of the organism. The phenotype, including the behaviour, is a product of these biochemical pathways. And behaviour being the synthesis of muscular, nervous, metabolic and sensory events, is necessarily influenced by numerous biochemical pathways. Given the enormous range of factors that could intervene between a gene and any influence it has on behaviour, it is all the more remarkable that it is sometimes possible to chart the influence of a particular gene on a particular pattern of behaviour, as in the case of the hygienic and non-hygienic bees.

3.6.3 Masking genetic effects

An individual who fails to produce the enzyme phenylalanine hydroxylase and who therefore has the mutant allele in their genotype, need not suffer the secondary symptoms of phenylketonuria. By the relatively simple expedient of reducing the amount of phenylalanine in the diet, an individual with PKU can grow and develop fairly normally. The blood test given to babies at around seven days of life is used to detect high levels of phenylalanine, and if necessary, a corrective diet is given. Thus the consequences of having a particular allele in the genotype can be overridden by a particular set of environmental circumstances. Indeed, in a world with no foods rich in phenylalanine, phenylketonurics would be indistinguishable from other people. In a world consisting entirely of phenylketonurics, meat, with its high content of phenylalanine, would be regarded as a harmful food!

The behavioural influences of particular genotypes can be similarly altered by changes in the environment in which the organism grows and develops. Earlier you read about Tryon's selection experiment in which he established two strains of rat: maze-bright and maze-dull. The differences in performance of these two strains in a particular type of maze were very clear, and as the differences were sustained from generation to generation, a reasonable deduction is that the differences had a genetic component. Tryon's strains were used by a number of other investigators to examine how robust the differences in the strains were. For example, R. Cooper and J. Zubek (1958) wondered whether the environment in which the rats were kept from weaning to early adulthood (i.e. from 25 to 65 days of age) would influence their performance in the maze learning task.

Rats are normally raised in a cage containing maybe a few other rats, some food, some water and not much else. With the addition of a few toys (ramps, mirrors, marbles, barriers, tunnels, see-saws etc.) these environments can be 'enriched'—though still of course impoverished when compared to the environment of the rat in the wild. Cooper and Zubek compared the maze learning performance of maze-bright and maze-dull rats raised in cages with toys ('enriched') and without toys ('normal'). The mazes that they used to test the rats were comparable with those used by Tryon, in that when rats encountered a dead end, of which there were many, an error was registered automatically. Thus performance in the maze was measured in terms of the cumulative number of errors a rat made in several trials of moving from the start box to the goal box: zero errors would be a perfect performance. The results are presented in Table 3.3.

Table 3.3   Mean error scores for maze-bright and maze-dull rats reared in different environments.

	Enriched environment	Normal environment
Bright	111	117
Dull	119	164

☐   Did the environment affect the rats' performance on the maze learning task?

■   The maze performance of the maze-dull rats which experienced the enriched environment was statistically indistinguishable from the maze performance of the maze-bright rats which experienced the enriched environment. The maze performance of the maze-dull rats in the normal environment was, however, not as good as the maze performance of the maze-bright rats in the normal environment (i.e. the results for this group were similar to those found by Tryon).

Clearly the experience of the rats affected their performance on this measure of behaviour. Any behavioural differences between these strains of rat that might be attributed to genetic differences could be obscured by raising the rats in different environments.

3.6.4 Phenocopies

There are numerous examples where altering the environment of an organism during its development affects the ultimate phenotype of the organism to such an extent that it no longer appears to be wild-type. In other words it appears to have a particular mutation. Such organisms are called **phenocopies**. For instance, there are three mutants of *Drosophila* that have broad wings, narrow wings and dumpy wings respectively. Normal flies, flies without any of those mutant alleles in their genotypes, can produce broad, narrow or dumpy wings, if subjected to a brief period of high temperature at particular times during their development. Many other phenocopies of *Drosophila* have been produced, including yellow.

Many vertebrates have two distinguishable sexes, male and female, and two distinguishable sex chromosomes, usually referred to as the X and the Y chromosomes. The sexual phenotype of an individual, however, how they look and behave, also referred to as their **gender**, may be at odds with their genetic sex. Thus a male phenotype may result from a female genotype. In cases where an individual's sex is in dispute (e.g. in athletics) it is usually the genetic sex of an individual (i.e. their chromosomal composition) that is regarded as definitive.

Some vertebrates have two sexes, but no sex chromosomes. In the tortoise (*Testudo graeca*), for example, it is not possible to distinguish the chromosomes of a male from those of a female. Somehow this tortoise is able to mimic the usual effects of sex chromosomes in the absence of sex chromosomes. What is crucially important in this species is the temperature of the eggs. At 28 °C, *Testudo* eggs develop into males, whereas at 32 °C they develop into females. Other species, e.g. many fish, are even able to change sex during their lives, being male up to a certain size, and then becoming female, or vice versa.

These examples illustrate, first, that environmental factors operating during development, including heat and cold, can affect the formation of morphological characters (e.g. wings). And, second, that in some species differences between the sexes are the result of environmental differences rather than genetic differences.

Summary of Chapter 3

This chapter has examined the relationship between genetics and behaviour. The genetic material, DNA, which is passed from adult to offspring in the form of chromosomes in the gametes, contains sufficient information to make thousands of different proteins. The proteins control the functions and structure of the cells. A change in the DNA, a mutation, can result in an altered protein which can affect the functioning of the organism.

Sometimes it is possible to trace differences in behaviour to individual genes, as in the case of hygienic bees; more usually, differences in behaviour can only be related to differences in genotype. A difference in genotype between animals does not necessarily mean that those animals will exhibit a difference in behaviour. A genetic difference may be masked by an environmental effect, or a genetic difference may simply alter the rate at which a behaviour pattern is performed, or the probability of a behaviour pattern being performed.

Genes do not cause patterns of behaviour directly, but they do influence them.

Objectives for Chapter 3

When you have completed this chapter, you should be able to:

3.1   Define and use, or recognize definitions and applications of each of the terms printed in **bold** in the text. (*Questions 3.1, 3.2 and 3.8*)

3.2   Give a brief account of basic genetics. (*Questions 3.3 and 3.4*)

3.3   Recognize false assertions about the relationship between genes and behaviour. (*Question 3.5*)

3.4   Give examples of single gene effects. (*Question 3.6*)

3.5   Give examples of polygenic effects. (*Question 3.6*)

3.6   Explain how the different techniques used in behaviour genetics can establish the link between the genotype and behaviour. (*Questions 3.7 and 3.8*)

3.7   Give examples of the way in which the environment may influence the expression of the genotype. (*Question 3.8*)

Questions for Chapter 3

Question 3.1 (*Objective 3.1*)
Distinguish between the following terms: locus, allele, gene, homozygous, genotype.

Question 3.2 (*Objective 3.1*)
Select from the list (1–5) the appropriate term(s) to complete sentences (a) and (b).

1   a mutation

2   a protein

3   a gene

4   an allele

5   a zygote

Sentence (a): A sequence of DNA bases is ....

Sentence (b): A sequence of DNA bases codes for ....

Question 3.3 (*Objective 3.2*)
Why is it only half correct to say that chromosomes are passed on from one generation to the next?

Question 3.4 (*Objective 3.2*)
How many different forms can a single gene come in?

Question 3.5 (*Objective 3.3*)
At the behavioural level the Tfm allele is associated with low levels of aggression. Does this mean that the normal allele at the same locus as the Tfm allele is the gene causing aggression? Explain your answer.

Question 3.6 (*Objectives 3.4 and 3.5*)
In *Drosophila* the yellow allele affects behaviour and also body colour. Is this an example of pleiotropy or polygeny?

Question 3.7 (*Objective 3.6*)
Several different styles of prey killing are seen in cats and the styles are thought to have a genetic component. Give three ways in which you could investigate the genetic component of prey killing in cats.

Question 3.8 (*Objectives 3.1, 3. 6 and 3.7*)
The garter snake (*Thamnophis sirtalis*) gives birth to live young (it is viviparous). These young are immediately able to strike out at any appropriate prey item. In one population the young eat slugs but, in another population the slugs are ignored. The offspring of laboratory crosses between these two populations ignore slugs. Give three possible reasons for the behaviour of the hybrids.

References

Bastock, M. (1956) A gene mutation which changes a behaviour pattern, *Evolution*, **10**, pp. 421–439.

Benzer, S. (1973) Genetic dissection of behaviour, *Scientific American*, **229**, pp. 24–37.

Cooper, R. M. and Zubek, J. P. (1958) Effects of enriched and restricted early environments on the learning ability of bright and dull rats, *Canadian Journal of Psychology*, **12**, pp. 159–164.

Dilger, W. C. (1962) The behaviour of lovebirds, *Scientific American*, **206**, pp. 88–98.

Hahn, M. E. and Haber, S. B. (1982) The inheritance of agonistic behavior in male mice: a diallel analysis, *Aggressive Behavior*, **8**, pp. 19–38.

Rothenbuhler, W. (1964) Behaviour genetics of nest cleaning in honey bees, *American Zoologist*, **4**, pp. 111–123.

Tryon, R. C. (1940) Genetic differences in maze-learning ability in rats, *Yearbook of the National Society for the Study of Education*, **39**, pp. 111–119.

CHAPTER 4
BEHAVIOUR AND ADAPTATION

4.1 Introduction

From previous chapters you will know that different animals within a species often behave in different ways. This variation can be due to genetic differences between individuals, to their having developed in different environments, to changes in their motivation, and so on. An individual animal may behave differently at different times and in different places. In this chapter you will learn about some of the ways in which animals vary and how studying this variation can provide insight into the benefits to be gained by behaving in one way rather than another. This chapter thus discusses the aspect of behaviour, introduced in Chapter 1, that is called its *function*.

4.2 Variation

4.2.1 Individual variation

Each individual consists of a collection of characteristics, or *characters*. These characters will vary. No two individuals of the same species are ever *exactly* the same—even genetically identical human twins brought up under similar conditions differ slightly in appearance, personality and behaviour.

Variation may be continuous or discrete. Where a character shows **continuous variation**, it may take any value, between certain limits, and, using that character, individuals cannot be divided into clearly distinguishable classes. Many characters show continuous variation—examples include size, colour, metabolic rate and fighting ability, to name but a few. For example, people could probably be found with heights of 1.5 m, 1.52 m or 1.500 012 34 m, depending on how accurately they are measured. For each pair of individuals it is possible to find another individual whose height is intermediate between them.

Where a character shows **discrete variation** (also called **discontinuous variation**), individuals *can* be divided into clearly distinguishable categories, between which there are no intermediates. People vary discretely in their blood groups. There are several human blood groups—O, A, B and AB, for example—and each person has blood of one particular type. There are no intermediates—one cannot have a blood group halfway between, say, O and A.

Where two or more discrete forms of a character are found within a single **population**, that character is said to be **polymorphic**. The term population refers to any group of individuals from the same species from a particular area, e.g. all the blackbirds (*Turdus merula*) living in a particular wood, or, on a larger scale, all the blackbirds in Britain. Individuals in a population may have something else in common; the term can also be used to describe, for example, all the red-haired people living in Bradford or all the blackbirds in a wood that are over two years old.

The most obvious example of polymorphism is sex—males and females are clearly distinct types within a single population, with generally no intermediate forms. There are very many other examples. During the breeding season, male ruffs (*Philomachus pugnax*), a wading bird, are of two types: territorial and satellite males (Figure 4.1). Ruffs have communal mating areas known as leks, in which some males maintain small areas, called territories, where they display to females. Territorial males fight with other males to keep them out of their territory so that they can mate undisturbed with females that come to the lek to mate. Satellite males do not fight or defend a territory. Instead they wait near a territorial male and achieve copulations while he is otherwise occupied! Each male is either territorial or a satellite and there are no intermediates. Even within discrete types, however, there can be *additional* variation. Territorial male ruffs vary continuously in how much fighting they do, for example, or how effectively they display to females.

Figure 4.1 Ruffs displaying at a lek. Males with dark ruffs are territorial; the male with a white ruff is a satellite; the bird with no ruff is a female.

Some of this variation between individuals is caused by differences in genotype, some by differences in the **environment** in which the individuals developed. 'Environment' in this sense consists of everything other than the genotype, and includes not only the physical world outside and all the other living organisms that surround the individual and with which it interacts directly or indirectly but also the internal environment within the individual itself. An individual's phenotype is the result of continuous interaction during development and adulthood between its genotype and its environment. This will be discussed in more detail in Chapter 5.

As far as territorial and satellite male ruffs are concerned, it seems likely that this variation is, at least in part, based on genetic differences between males—they have distinctly different plumages associated with their behavioural differences, and males never apparently change from one type to the other.

4.2.2 Geographical variation

In some species, as well as individual variation, a considerable amount of variation is also found between populations in different parts of the geographical area inhabited by the species (its **range**).

Where distinct types are found in different parts of a species' range, that species is said to be **polytypic**. Depending on the degree of difference between them, these distinct types can be designated as races (slight differences) or subspecies (greater differences). Subspecies are designated by the addition of a third part to the Latin name of the species. The crow, *Corvus corone*, for example, has two subspecies— the all-black carrion crow, *Corvus corone corone*, which inhabits western Europe including southern Scotland and eastern Ireland, and the hooded crow, *C. c. cornix*, which is grey with a black head, wings and tail and inhabits central and eastern Europe, northern Scotland and western Ireland (see Figure 4.2).

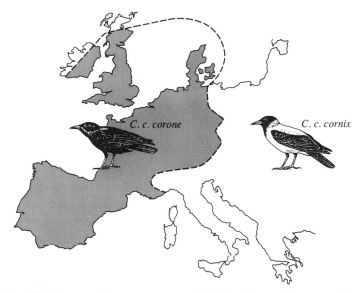

Figure 4.2 Distribution of the carrion crow and the hooded crow in western Europe.

4.3 Natural selection and evolution

4.3.1 Reproductive potential

Usually, each individual in a species has the potential to produce more offspring than is necessary to replace itself. Numbers of a sexually reproducing species, for example, will remain constant if each mating pair, during its whole lifetime, produces only two offspring that survive long enough to breed themselves. Most individuals are capable of producing many more offspring than this. One of the slowest breeding animals is the African elephant (*Loxodonta africana*). It does not reach sexual maturity until it is about 10 years old, has a gestation period of 22 months and suckles its single calf for 3–4 years.

☐ A single pair of elephants mate when aged 10 years and the female gives birth to their first calf 22 months later. Thereafter they have another calf every 5 years. How many calves will they have had by the time they are 33 years old?

■ 5. They will produce a calf when they are approximately 12, 17, 22, 27 and 32 years old.

If all the calves survive to adulthood and are fertile, they will in turn start producing calves once they reach sexual maturity. By the time the parents are 33 years old, their first calf would be 21 years old and, if it happened to be a female, could have given birth to another 2 calves itself. Similarly, the parents' second calf would be 16 years old and, if it were a female, could have had one calf. So, by the time the original pair are 33 years old, there might be as many as 10 elephants, a five-fold increase in numbers in only 23 years.

Thus numbers of even a very slow-breeding species could increase substantially in a relatively short space of time if every individual survived to breed. A similar calculation for a fast-breeding species like the European rabbit (*Oryctolagus cuniculus*), which reaches sexual maturity at 3–5 months, and can produce litters of about 6 young every 2 months from February to August for up to 10 years of life, would give a figure of well over 4 million rabbits after only 5 years! For a species like the cod (*Gadus callarias*), which can lay 3–4 million eggs per year, such calculations quickly result in astronomical numbers.

Obviously, no species can be reproducing at its full potential. Most species stay at relatively constant numbers simply because many young animals and plants die before they reach maturity and, of those that do survive long enough to reproduce, many do not have the maximum number of offspring that, in theory, they are capable of producing. Individuals die because of accidents, unfavourable environmental conditions, disease, predators or a shortage of essential resources such as food. If they survive to breeding age they may be unable to obtain a mate or may suffer from reduced fertility (see Section 4.3.3) as a result of genetic factors, disease, shortage of essential resources and so on.

For example, although European rabbits can live up to about 10 years in the wild, the *average* lifespan is less than 1 year and most young die before they ever get to breed. The number of young produced by an adult female also varies enormously—some females are capable of having litters of 12 young five times a year, others breed only three times a year and have less than half that number of young.

4.3.2 Competition

The number of offspring produced each year by a female rabbit depends to a large extent on the amount of food available to her. An important factor here is that she has to compete with other rabbits for food.

☐ What three main factors will determine how much food a female rabbit obtains in a situation where she is competing with other rabbits?

■ How much food is available, how many other rabbits she has to compete with and how good she is at competing with other rabbits.

The term **competition** is used here to cover both direct and indirect competition. **Direct competition** involves direct physical interaction with competitors, **indirect competition** does not. For example physically blocking other rabbits' access to a particular resource or actually fighting with them to obtain it would be direct competition. Being better at finding food than other rabbits, or having a more efficient digestive system so that more nourishment is gained from any food that is obtained, would be indirect competition. Whatever form the competition takes, ability to compete is a key factor determining not only which individuals will survive to breed, but also how many offspring they will leave if they do reproduce.

Because the resources of the environment are limited, all living things have to compete with each other for those resources, either directly or indirectly, and not just for food and water but also for space, shelter, oxygen, mates and so on. Even where individuals apparently cooperate, such as a group of hyaenas acting together to kill a prey animal, this is probably because each individual does better by such cooperation. Hyaenas hunting together have to share the meat they obtain but they are more likely to have a successful hunt and can kill much larger animals than a hyaena acting alone. Cooperation will be discussed in Section 4.3.7 of this chapter and in Chapter 10.

4.3.3   Natural selection

Individuals possessing a characteristic that enables them to compete successfully for a limited resource will be more likely to survive and reproduce than others less able to compete. They will therefore tend to leave more offspring. If the characteristic is subject to genetic control, it will tend to be inherited by the offspring. Therefore, as long as environmental conditions stay the same, the offspring, in their turn, will also be more successful and will leave more offspring than members of the population without the characteristic. The characteristic will thus tend to spread through the population—i.e. more individuals in the population will possess it—and any allele affecting the development of the characteristic will increase in frequency. Conversely, less successful phenotypes will tend to decrease in numbers and alleles affecting the development of such phenotypes will decrease in frequency. This process of differential success in surviving and reproducing, by which the number of individuals possessing a particular characteristic changes over the course of one or more generations, is called **natural selection**. More successful phenotypes and the alleles contributing to their development are said to be selected *for*; less successful versions are said to be selected *against*.

Natural selection can only act when there is variation, since if all individuals are the same, and they live in identical conditions, no individual can have any particular advantage over the others. It can only act on characteristics that have a genetic component, since only these characteristics can be inherited by an individual's offspring. On the other hand, it can only act on the phenotypic characteristics of individuals, rather than directly on their genotype, because it is the end result of the interaction between genotype and environment, i.e. the phenotype, that affects the individual's actual ability to survive and reproduce. The end result, however, is a change in the frequency of a particular allele or set of alleles in a population over a number of generations. It is also important to remember that natural selection has no 'foresight'; it cannot predict how the environment will change, but only operates on variation in a population present at a given time, in the environment that prevails at that time. Evolutionary change by natural selection thus typically follows changes in the environment.

4.3.4 Fitness

In everyday usage, *fitness* generally means good health or good condition. As a biological term, however, it means something much more specific. **Fitness** is a measure of **lifetime reproductive success** *relative* to other members of the population (and as such is often called **relative fitness**). The lifetime reproductive success of an individual is the number of fertile offspring it produces during its lifetime that survive to breed themselves. So an individual whose lifetime reproductive success is greater than that of other members of the population is said to have greater fitness (or to be more fit).

Every individual, however, possesses many characteristics, some of which may increase fitness and some of which may lower fitness. Overall fitness will be the sum total of all these **components of fitness**. A particular vixen for example may be a very skilful hunter, which will tend to increase her fitness, but may be particularly susceptible to certain diseases, which will tend to decrease her fitness.

There are obviously very many components of fitness but they can be divided into two main categories:

1   Those characters that directly affect lifetime reproductive success, such as the ability to find a safe place to lay eggs or to choose a suitable mate.

2   Those characters, such as greater resistance to disease or extra speed to outrun predators, that increase lifetime reproductive success *indirectly* by increasing the chance of survival, so that an individual is more likely to reach breeding age and, in a species that breeds more than once, may have more chances to breed.

Lifetime reproductive success is directly dependent on a number of factors:

1   Fecundity—the number of gametes, either eggs or sperm, produced.

2   Fertility—the proportion of eggs produced that are capable of being fertilized or the proportion of sperm cells produced that are capable of fertilizing an egg.

3   Viability of offspring—the proportion of zygotes that complete normal development and survive to reach breeding age themselves.

Because fitness is always measured relative to other members of the population, a fitness of e.g. 0.82 for a particular phenotype means that this phenotype is only 82% as successful as the most successful phenotype in the population (which has a relative fitness of 1.0). The **selection pressure** (the strength of selection) against an inferior phenotype will depend on how successful it is compared with the most successful phenotype in the population. The selection pressure can thus be expressed as the difference between the relative fitness of the best phenotype (1.0) and the relative fitness of the inferior phenotype (0.82 in the example given above), i.e. $1.0 - 0.82 = 0.18$. This value—0.18 in this example—is called the **selection coefficient**.

You can see from this that natural selection is not an all-or-nothing process. A phenotype selected against does not inevitably die young. The strength of selection varies. Even very weak selection leading to a very small average difference between two kinds of phenotype in lifetime reproductive success will, if the selection pressure is maintained in the face of environmental change, result in a gradual increase in the more successful phenotype, even though it may take many generations before the more successful phenotype replaces the other.

Tim Clutton-Brock and his colleagues have been studying red deer (*Cervus elaphus*) on the Isle of Rhum off Scotland (Figure 4.3) for many years (Clutton-Brock *et al.*, 1982). They have estimated the lifetime reproductive success for females (Figure 4.4). The female that did best in the course of this study reared 13 surviving calves during her life.

Figure 4.3   A female red deer and her day-old calf on the Isle of Rhum.

☐   What is the relative fitness of this female?

■   Because this female did best, her relative fitness must be 1.0.

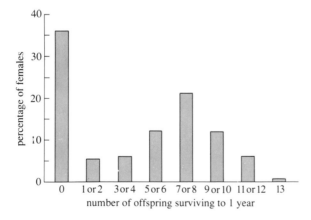

Figure 4.4   Estimated lifetime reproductive success for females in a population of red deer on the Isle of Rhum, calculated as the number of offspring surviving to 1 year.

Other females reared from 0 to 12 calves, and the average for the whole population was 4.5 calves. An average female would have a relative fitness of 4.5/13 = 0.35, i.e. she was only 35% as successful as the female that did best.

☐ What is the selection coefficient for an average female?

■ 1.0 − 0.35 = 0.65.

4.3.5 An example of natural selection in an imaginary population

Consider an imaginary population of small mammals consisting of 50 breeding pairs. There is a short annual breeding season. Individuals mate only once, produce 8 young per pair, and then die. Only 25% of the young born in one year survive to breed the next.

☐ How many individuals will there be in the population at the start of the breeding season 2 years later?

■ 100. The 50 breeding pairs have 8 young each, a total of 400, but only 25% of these survive, leaving 100. The number of adults at the start of the breeding season each year therefore remains constant.

In other words—and as you will remember from Section 4.3.1—the numbers of a species will remain constant as long as each mating pair on average produces only two offspring that survive long enough to breed themselves.

Now suppose that an individual is born in this imaginary population that has a mutation in one of its genes. It has a 'new' allele at a particular locus, whereas all other members of the population have two copies of the original, unmutated allele. This genetic difference affects the mutant's phenotype such that its chance of surviving to breed is twice that of other members of the population. In other words it is twice as *fit* as other individuals. It might be twice as good at avoiding predators, for example, or better able to compete for food.

For convenience, imagine that this individual is better able to compete for food. Let us call it a 'Grabber' and the mutant allele the Grabber allele (*G*); the other members of the population are 'non-Grabbers', with the original, non-Grabber allele (*g*). Let us assume also that:

1 A Grabber has a 50% chance of survival instead of the 25% chance that non-Grabbers have.

2 The first Grabber survives to breed.

3 The Grabber allele is dominant and if a heterozygous Grabber mates with a non-Grabber, half of its offspring will inherit its ability to compete for food and so will also be Grabbers (its other offspring will be non-Grabbers).

4 That the size of the adult population does not necessarily have to remain constant at 100.

☐ Given this, how many Grabbers are there likely to be at the start of the breeding season in the year after the original Grabber dies?

■ Two. The Grabber will have 8 offspring, of which 4 will be Grabbers since it will have to mate with a non-Grabber. These will each stand a 50% chance of surviving so, on average, 2 out of the 4 Grabber offspring will live until the next breeding season.

The number of Grabbers has thus doubled in the space of 1 year.

☐   How many individuals will there be in the population at this point?

■   101. 25% of the 392 (= 49 × 8) offspring from the 49 non-Grabber pairs will survive, i.e. 98. 25% of the 4 non-Grabber offspring of the Grabber will also survive, i.e. 1. You have also just calculated that 2 Grabber offspring will survive, giving a total of 98 + 1 + 2 = 101 individuals in the population.

☐   What do you notice about the number of non-Grabbers in the population?

■   It stays the same—99.

If you assume that a Grabber always mates with a non-Grabber, then the number of Grabbers will double each year. After only 7 years, there will be 128 Grabbers—more than half of the population because there will still be only 99 non-Grabbers. If, on the other hand, you assume that mating is random among Grabbers and non-Grabbers then Grabbers will still increase—the increase will just be slower.

☐   How would the rate of increase of Grabbers in the population over the first 7 generations be affected if the difference in relative fitness between Grabbers and non-Grabbers (i.e. the selection pressure on non-Grabbers) were less than in the example given above?

■   The rate of increase would be slower because Grabbers would have less of an advantage over non-Grabbers.

Since most natural populations tend to stay at relatively constant numbers, the above example would be more realistic if the number of individuals in the population stayed constant rather than increasing.

☐   If you did the above calculations again, but this time assumed that adult numbers in the population stayed constant at 100, what would tend to happen to the number of non-Grabbers compared with Grabbers each generation?

■   The number of non-Grabbers would decrease as the number of Grabbers increased.

For numbers to stay at 100, only 25% of all offspring can survive overall. If Grabbers are always twice as likely to survive as non-Grabbers, the exact figures for the first generation (i.e. the generation that includes the original Grabber's offspring) will be 49.5% of Grabbers and 24.75% of non-Grabbers survive until the next breeding season.

This imaginary example is a very simplistic one because, in reality, individuals differ genetically at very many loci, not just one, and both environmental and genetic factors contribute to their overall fitness. Having a different allele at just one locus would be very unlikely to produce such an enormous difference in fitness as that between the hypothetical Grabbers and non-Grabbers. But the example illustrates how natural selection, by acting on *individuals*, can cause an advantageous phenotype (as long as it is inherited) to spread through a *population*. Grabbers are selected for and non-Grabbers are selected against and, eventually, the population would consist almost entirely of Grabbers. But, since the advantage of being a Grabber is that it is twice as likely to survive as a non-Grabber, i.e. fitness is measured relative to others rather than having an absolute value, the benefit an individual gains from being a Grabber depends on how many *non-Grabbers* there are around.

☐ Suppose that a particular population consists of 99 Grabbers and 1 non-Grabber. Assuming that the number of adults in the next generation stays at 100, will the proportion of the Grabbers' offspring that reach breeding age be: (a) approximately 50%; (b) 49.5% or (c) approximately 25%?

■ (c) approximately 25%. When the population consists almost entirely of Grabbers, nearly 75% of their offspring must die if the number of adults at the start of the next breeding season is to remain at 100.

☐ Remember that 49.5% of the original Grabber's offspring reach breeding age if the number of adults in each generation stays at 100. What do you conclude about the chance of surviving for an individual Grabber as the proportion of Grabbers in the population increases?

■ Its *actual* chance of surviving decreases though, by definition, it is always twice that of a non-Grabber.

Suppose that eventually the whole population came to consist of Grabbers homozygous for the *G* allele. As the only type present, Grabbers and the *G* allele are then said to have become **fixed** in the population. Non-Grabbers and the *g* allele have become extinct. In this situation, there would no longer be any advantage in being a Grabber, in terms of the individual's own survival. In other words, the selection pressure on the non-Grabber changes as the percentage of Grabbers in the population changes, becoming less and less as the proportion of Grabbers increases. Selection pressure on a particular phenotype will vary with *any* changes in the environment—which you will remember includes other members of the same species.

Variable selection pressures as a result of a variable environment can lead to polymorphisms or polytypies. For example, if for some reason Grabbers were favoured by selection at some times and non-Grabbers at other times, then neither type would be likely to become fixed and the population would be polymorphic. Similarly, if Grabbers were fitter than non-Grabbers when there were only a few Grabbers in the population while non-Grabbers were fitter than Grabbers when there were only a few non-Grabbers in the population, each type would increase in numbers when it was in a minority and the result would be a polymorphic population consisting of both Grabbers and non-Grabbers. The environment will vary in different parts of a species' range and if, for example, Grabbers were selected in one area while non-Grabbers were selected in another area, the result would be a polytypy.

One example of a polymorphism that depends on the relative abundance of the different types is provided by the ruff, mentioned in Section 4.2.1. We will return to this example later in this chapter (Section 4.4.4).

Summary of Section 4.3.5

Alleles that contribute to the development of a phenotype with greater than average fitness will spread through a population at the expense of alleles that contribute to the development of less fit phenotypes. As long as environmental conditions remain unchanged, such alleles may continue to spread until they become fixed. The greater the relative fitness of the phenotype, the faster the alleles contributing to its development will spread. But the spread of the fitter phenotype may itself cause the environment to change, so the relative fitnesses of different phenotypes may also change, leading to the establishment of polymorphisms or polytypies.

4.3.6 Types of natural selection

Evolution can be defined in a number of ways. For the purposes of this course, it will be defined as a change over a number of generations in the frequency within a population of a particular phenotype or of one or more alleles contributing to the development of that phenotype. Such a change does not necessarily have to be the result of natural selection. Natural selection can account for changes in species if there is *existing* variation; it does not account for the *origin* of that variation, for example by mutation. Chance can lead to evolution. For example, a particular characteristic might be passed on to more or fewer offspring than expected by chance, or more or fewer individuals with that characteristic might survive than expected by chance (Figure 4.5, *overleaf*). Such chance fluctuations in phenotype or allele frequency are called **genetic drift**. Genetic drift can be an important factor in small populations, where a small increase in the *number* of individuals with a particular characteristic results in a large increase in the *percentage* of those individuals in the population.

Conversely, even if selection is operating, evolution may not be taking place. For example, selection against a particular phenotype may remove individuals with that phenotype from a population at the same rate as new individuals with that phenotype arise by mutation or recombination, thus maintaining a balance so that the frequency of the particular phenotype does not change. Selection that acts in such a way that the proportions of each phenotype remain constant is called **stabilizing selection**.

Stabilizing selection may operate in another way. Imagine a population of crickets. Males of this imaginary species call to attract mates. Chirp rate shows continuous variation: adult males vary in the rate at which they call, from 20 to 30 chirps per min (cpm), with the average rate being 25 cpm. The distribution of male chirp rates is shown in Figure 4.6a (*overleaf*).

Suppose that a scientist is able to capture the whole population of crickets and take them into the laboratory. For breeding purposes, the population is divided up into separate pairs consisting of a single male plus a single female. Individuals are reassigned randomly to pairs at the start of each breeding season. All individuals thus have equal opportunities to breed and mating is random with respect to chirp rate. After several generations in the laboratory, the scientist measures male chirp rate again and finds that the distribution is now that shown in Figure 4.6b.

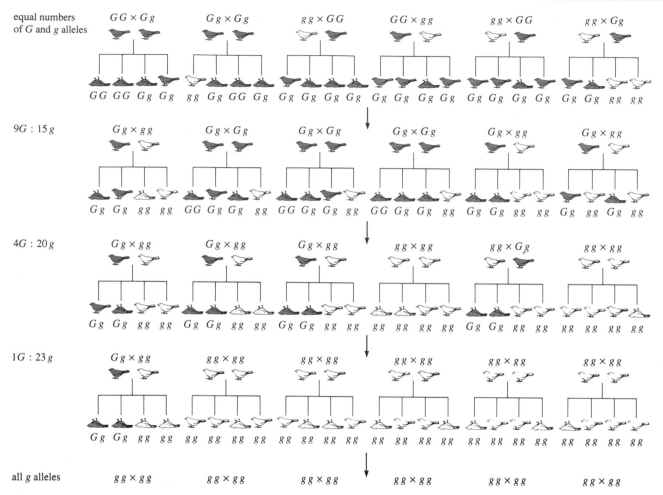

equal numbers of G and g alleles

9G : 15g

4G : 20g

1G : 23g

all g alleles

Figure 4.5 Genetic drift in a small population of birds. There are two alleles for plumage colour, black (G) and white (g). Black is dominant. Black and white phenotypes are equally fit. In each generation, mating is random and each pair produces 4 offspring. Half the young die and the population remains constant in size. Which particular young die is, however, entirely random and not dependent on genotype or phenotype. The frequency of each genotype fluctuates at random until, by chance, the black allele becomes extinct and white becomes the only allele in the population.

☐ Compare Figure 4.6b with Figure 4.6a. What do you conclude?

■ The range of chirp rates is greater in the laboratory population. The average chirp rate is the same as in the wild but the laboratory population contains some males that chirp at a slower rate and others that chirp at a faster rate than any males found in the wild.

The scientist carries out a field study of another population of the same cricket species, similar to the original population, and discovers that, among male crickets, individuals that chirp slowly are less likely to attract mates, so that they leave fewer offspring, while individuals that chirp quickly are attacked preferentially by predators (Figure 4.6c). As a result, the adult population contains fewer slow and fast chirping individuals than might be expected and the fittest individuals are those that chirp at an average rate. Stabilizing selection in this case keeps the average chirp rate the same and reduces the overall variation in rate;

over time, the amount of variation and the range of chirp rates stays pretty much the same. There is no evolution as far as chirp rate is concerned.

Selection only leads to evolution if the proportions of different phenotypes in the population change.

☐   In the wild cricket population above, what would happen to the range of chirp rates if predators chose their prey non-selectively and regardless of the rate at which they were chirping?

■   There would be no selection operating against faster-chirping individuals, though selection against slower-chirping individuals as a result of their leaving fewer offspring would continue to operate. As a result, both the range of chirp rates and the average chirp rate in the population would increase (see Figures 4.7a and 4.7b).

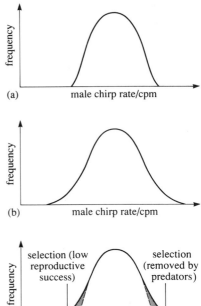

Figure 4.6   Distribution of male chirp rates (chirps per minute or cpm) for an imaginary population of crickets: (a) the population in the wild; (b) the population after several generations spent in the laboratory; (c) the action of stabilizing selection on the wild population.

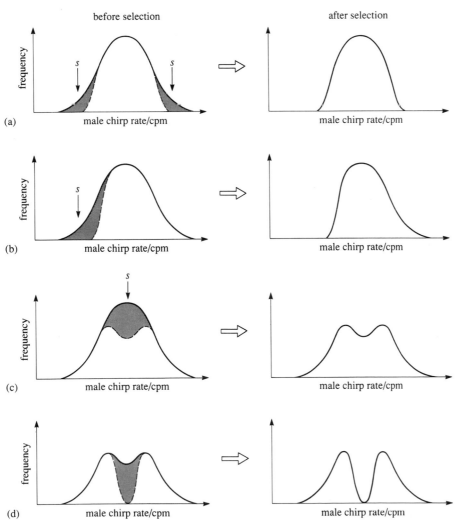

Figure 4.7   Selection acting on male chirp rate in a hypothetical wild population of crickets: (a) the population under stabilizing selection; (b) the population under directional selection; (c) the population under disruptive selection; (d) disruptive selection may lead to the establishment of a polymorphism for chirp rate.

This is an example of **directional selection**. More slow chirpers than fast chirpers would be removed from the population and the proportion of fast chirpers in the population would increase while that of slow chirpers would decrease. An evolutionary change would have taken place and crickets in this population would on average chirp faster.

Taking the same wild population of crickets, imagine that females are equally attracted to males chirping at any rate; imagine also that environmental conditions change so that the climate gets cooler. As a result, the main predator on the population (a species that can only survive in a warm climate!) disappears and is replaced by one that mainly catches crickets that chirp at a medium rate, perhaps because it can more easily locate such individuals in the vegetation.

☐   Can you predict what will happen to the distribution of chirp rates in the population?

■   Individuals that chirp at a medium rate will be taken by the new predator and so will be less fit. They will become rarer relative to those that chirp faster or more slowly (see Figures 4.7c and 4.7d).

In this case, slow-chirping and fast-chirping individuals will both be favoured by selection at the expense of those that chirp at the average rate. Evolution takes place as the result of such **disruptive selection** because the proportions of fast, medium and slow chirpers in the population change. Disruptive selection can, under certain conditions, lead to the establishment of a polymorphism, with the population consisting only of individuals of the two extreme kinds, with no intermediates (Figure 4.7d), or even two separate species if the two types diverge to such an extent that they can no longer interbreed.

4.3.7 Inclusive fitness and kin selection

Many group-living bird and mammal species give 'alarm' calls in response to potential danger, usually after a predator has been spotted. There is a real cost to the caller in performing this behaviour—the increased likelihood of attracting the predator's attention to itself—though the cost is minimized because alarm calls are often high-pitched and difficult to locate. A behaviour pattern like this, that benefits another individual at a cost to the performer's own individual fitness, is said to be altruistic. This definition is specific to the study of animal behaviour and you should note that it does not have the same meaning, or carry the same moral connotations, as when the word is used in everyday speech. Defined in this way, **altruism** does not include normal parental behaviour, since behaviour that increases an offspring's chances of surviving long enough to breed itself must, by definition, increase the parent's fitness. But how could altruism towards non-offspring evolve, since altruistic individuals should be less fit and therefore selected against?

In the past, it was often argued that altruistic or cooperative behaviour was 'for the good of the species'. In other words, the behaviour evolved because it benefited the group as a whole—a group in which individuals helped each other did better than one in which they did not. Thus cooperating groups were 'selected' at the expense of non-cooperating groups—non-cooperating groups were more likely to die out completely than cooperating groups, and 'offspring' of cooperating groups could then replace them, so cooperating groups tended to spread. This process is called **group selection**. But if you think carefully about this idea, you can see that it will not work. Even if, as seems unlikely, whole groups of individuals have completely died out often enough for group selection to have been an important factor in the *spread* of altruism, it does not explain how altruistic groups could have arisen in the first place.

Imagine a group that does *not* behave altruistically—individuals do not, for example, give alarm calls to warn other group members of danger. Then a mutant individual arises that *does* give alarm calls. This behaviour makes the altruist more likely to be caught by a predator (remember that, by definition there is always a fitness cost associated with altruistic behaviour), so the chances are it will not live as long as other group members and its lifetime reproductive success will be lower (it may not even breed). Thus altruists would be selected against and the altruist phenotype would eventually disappear from the population. Even if genetic drift operates and it is lucky enough to avoid predation, breed successfully and pass its 'altruist' alleles on to its offspring, the extra risks involved in giving alarm calls would, over the course of succeeding generations, reduce the average reproductive success of the altruists and their numbers would gradually decrease until they became extinct. For altruists to increase and become fixed in the population and non-altruists to become extinct, the altruists would have to do *better* on average than non-altruists, despite courting danger every time they gave an alarm call, or the population would have to be small enough for genetic drift to be an important factor.

It might still be possible for a group to evolve altruistic behaviour against all the odds, i.e. by genetic drift outweighing the effects of natural selection. Imagine such a group in which all individuals behave altruistically by giving alarm calls. Then a mutant individual arises that does *not* give alarm calls. As a result, it is more likely to evade predators.

☐   All other things being equal, is the lifetime reproductive success of this mutant individual likely to be (a) greater, or (b) less than other members of the group?

■   Greater, because it is less likely to attract the attention of a predator yet can still benefit from the alarm calls of other members of the group.

☐   Given this, should the numbers of non-altruists in the population tend to increase or decrease, and what should eventually happen to the altruists?

■   Non-altruists should increase and the altruists should eventually become extinct.

Thus, the non-altruist is always at an advantage over the altruist, and the altruists will always tend to become extinct. So, as long as altruistic behaviour is seen as *costly*, it is difficult to see how altruistic behaviour could have evolved.

In fact, the benefits to the altruist can outweigh the costs if its behaviour increases the fitness of its relatives. This is because it is not so much *individual* fitness that is important but **inclusive fitness**—the lifetime reproductive success of the individual concerned *plus* the lifetime reproductive success of related individuals that carry copies of the alleles that contribute to the development of the altruistic behaviour in question. Natural selection operating indirectly on a character by favouring the relatives of the individual having the character is called **kin selection**.

You will remember from Chapter 3 that a parent passes on only one chromosome from each pair to each of its offspring. There is therefore a 50 : 50 chance (i.e. a probability of 0.5) that an offspring will inherit a copy of a *specific* allele carried by its parent. Similarly, for full siblings (offspring of the same two parents) there is a 50 : 50 chance that they will *both* inherit a copy of a specific allele carried by one or other of their parents. This is because, given that one offspring inherits a particular allele from a parent, the chances of another offspring also inheriting the same allele are 50 : 50—it can only get one or other of the two alleles carried by that parent.

☐ If there is a 50% chance that a child will inherit a copy of a specific allele from its parent, what is the chance that a grandchild will inherit a copy of a specific allele from its grandparent?

■ 25% (in other words a 1 in 4 chance or a probability of 0.25). There is a 50% chance that a child will inherit a copy of a specific allele from its parent but, even if it does inherit it, there is only a 50% chance that it will pass on a copy of that same allele to its offspring. On average, therefore, only one in four grandchildren will inherit a copy of the specific allele in question.

Similarly, the chance that an individual and its half-sibling (only one parent in common), or an individual and its aunt, uncle, niece or nephew, will both inherit a copy of the same specific allele is 25%. The chance that first cousins will both inherit a copy of the same specific allele is 1 in 8 (12.5% or a probability of 0.125). This probability that two individuals will both inherit a copy of a specific allele from a common ancestor is called the **coefficient of relatedness** (r). For first cousins, r is equal to 0.125; for full siblings it is 0.5.

So when an individual shows altruistic behaviour to a relative, it increases the fitness of an individual likely to be carrying copies of the same 'altruist' alleles as itself. The altruist itself may experience a decrease in lifetime reproductive success, but the increased reproductive success of the individual it helps—an individual that is likely to carry copies of the same 'altruist' alleles—can compensate for this. Thus its own *inclusive* fitness may be increased. The end result is an increase in frequency of the 'altruist' alleles. The closer the relationship between the altruist and the individual it helps, or the greater the number of relatives helped, the greater the likelihood that an individual sharing the same 'altruist' alleles will benefit.

Social species (see Chapter 10) often live in groups of related individuals and several field studies have provided evidence that kin selection has been important in the evolution of altruistic behaviour. For example, P. W. Sherman (1977) studied alarm calling in a population of Belding's ground squirrel (*Spermophilus beldingi*), a burrowing animal living in social groups in the mountains of the western United States (Figure 4.8).

Females of this species tend to stay in the social group in which they are born and so live with females that are related to them, but males disperse and move from group to group between breeding seasons.

Figure 4.8   Belding's ground squirrel.

☐   If alarm calling evolved through kin selection, which do you predict should give more alarm calls, males or females?

■   Females, because they are related to other members of their social group while males tend not to be related. There is less advantage, therefore, in males warning other members of their group about approaching predators.

Sherman found that not only do females alarm call more than males but that females with relatives in addition to their own offspring nearby call more than females that only have their own offspring nearby.

Summary of Section 4.3.7

Many animals living in groups show altruistic behaviour that benefits others at a cost to their own individual fitness. Such behaviour could not have evolved simply because it benefits the group or the species as a whole. It could have evolved if it increases the reproductive success of relatives, thereby leading to an increase in the frequency of alleles contributing to the development of that particular altruistic behaviour. Many animal groups do consist of related individuals and there is evidence from field studies that this process of kin selection may have been important in the evolution of altruism.

4.3.8   Measuring survival and reproductive success

Many studies have attempted to measure the effects of particular characteristics on survival and lifetime reproductive success in wild populations. This is not always easy. For example, if the animal has a long lifespan, it will take many years to obtain values for lifetime reproductive success. On the other hand, if its lifespan is short, the animal is likely to be small and difficult to observe in the wild. It is often necessary, therefore, to measure reproductive success over part of the lifetime of an individual, but care must be taken in extrapolating such results. It cannot be assumed for example, that a male that has very few mates in one year will do equally badly the next. He may have been inexperienced, or suffering temporarily from injury or disease.

Populations of some long-lived species have been studied over considerable periods of time, however, so that reliable estimates of mortality and lifetime reproductive success are available. Data on estimated lifetime reproductive success for female red deer on the Isle of Rhum were given in Figure 4.4. These can be compared with those for male red deer from the same population, shown in Figure 4.9.

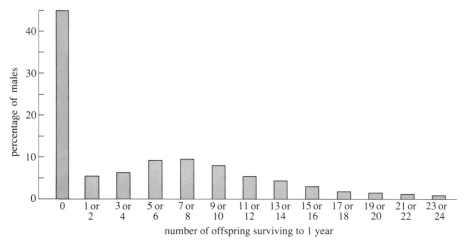

Figure 4.9 Estimated lifetime reproductive success for males in a population of red deer on the Isle of Rhum, calculated as the number of offspring surviving to 1 year.

☐ Compare Figure 4.9 with Figure 4.4. What do you conclude about the variation in reproductive success among females compared with that among males?

■ Variation in males is much greater (a range of 0–24) than among females (a range of 0–13).

Red deer are polygynous (i.e. each male may mate with several females—see Chapter 9) and during the breeding season, males compete with each other to gather a group of females with which they can mate exclusively (Figure 4.10). Females mate in most years of their adult life (they can live for up to about 20 years and usually breed from the age of three or four) and the variation among them in lifetime reproductive success is due mainly to differences in calf survival. Many males, however, never mate at all; others can father up to five calves in one season, though even the most successful males rarely breed for more than about four or five seasons. Variation in lifetime reproductive success among males is largely due to differences in fighting ability which in turn is related to size and condition. Any allele that contributes to the development of large body size in males, or to the ability to maintain good condition, will be favoured by natural selection. This does not mean to say that genes affecting characters other than fighting ability, size and condition are not important contributors to lifetime reproductive success. If he is to mate at all, a red deer stag must be recognizable

Figure 4.10   Red deer stag with a group of females during the breeding season.

as such to a red deer hind, for example, so that mating only takes place between conspecifics.

Edward McLean and Jean-Guy Godin (1989) in Canada measured the survival value of fleeing from a predator in three species of fish which varied in their amount of defensive body armour. Such armour provides a certain amount of protection against predation by larger fish because it makes the prey more difficult to swallow. The three-spined stickleback (*Gasterosteus aculeatus*) is heavily armoured, with three large dorsal spines and another two ventral spines, plus several lateral bony plates. The banded killifish (*Fundulus diaphanus*), on the other hand, has no armour at all, while the ten-spined stickleback (*Pungitius pungitius*) has an intermediate amount, with nine small dorsal spines, two small ventral spines and only a few lateral bony plates (see Figure 4.11, *overleaf*).

A fish hiding in vegetation is less likely than a fish in open water to encounter a predator. Even if it does, it is less likely to be caught because the vegetation will hinder the movements of the predator more than the movements of the prey, since a predator is usually considerably larger than its prey. On the other hand, for fish that feed on plankton, small animals or other fish, there is more food available in open water. There is thus a conflict between avoiding predators and obtaining food. A fish that encounters a predator in open water has two choices: flee to vegetation where it can hide, or stay where it is, perhaps remaining immobile in an attempt to escape detection.

☐   The benefit of fleeing is obviously the possibility that the predator is avoided, but what will the two main *costs* of fleeing be, and how will these vary as the distance to the nearest vegetation varies?

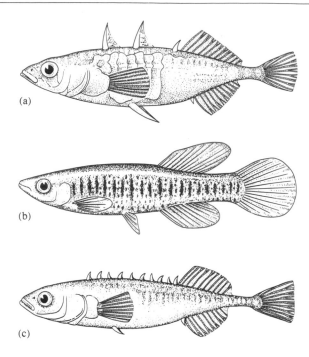

Figure 4.11 Body armour in three species of fish: (a) the three-spined stickleback; (b) the banded killifish; (c) the ten-spined stickleback.

■ The costs of fleeing will be the energy used to swim to vegetation and the lost feeding time while fleeing and hiding. These costs will increase as distance to nearest vegetation increases.

The closer the predator gets to its prey before it flees, the more likely it is to catch the prey. At the same time, the closer the fleeing prey is to vegetation, the more likely it is to reach cover before the predator can catch it. If the prey is a long way from vegetation, it needs longer to get there and so must flee from the predator earlier if it is to escape. If the prey is near to vegetation, it can remain immobile until the predator is relatively close because it needs only a short time to reach cover. The benefits of fleeing, should, therefore, increase as the distance to the nearest vegetation increases and the overall survival value of fleeing versus remaining immobile should vary depending on how far the fish is from cover. McLean and Godin set out to investigate these predictions in the laboratory by measuring fleeing tendency in response to a model predator for each of the three species of fish.

☐ Given all the information above, which of the three fish species would you predict as being the most likely to flee at the approach of a predator and which the most likely to stay where it is?

■ The banded killifish should show the greatest tendency to flee because it has the least amount of body armour so its risk of death through predation is greater. The three-spined stickleback should show the least tendency to flee because it has the greatest amount of body armour.

This is exactly what McLean and Godin found. They also observed, as predicted, that all fish showed a greater tendency to flee when they were a long way away from cover and so were at greater risk of being caught by the predator.

This study illustrates the fact that individual characteristics and behaviours are not selected in isolation. An individual that avoids predators is at an advantage compared with one that does not, but an individual that spends a lot of time avoiding predators at the expense of time spent foraging for food would be selected against, because its reproductive success is likely to be jeopardized. But why have not all three species of fish evolved body armour, since they all suffer from predation? Energy and resources are required to grow body armour—energy that could be put into reproduction, so here is another conflict, this time between avoiding predation and reproduction. This conflict has been resolved in different ways in different species, some species putting their energy into growing armour, others using their energy to flee.

4.4   Adaptation

For an animal, some of the consequences of a particular character may be beneficial, i.e. they tend to increase the animal's fitness, but some may incur costs, i.e. they tend to reduce fitness. If the total fitness benefits outweigh the total fitness costs, then the character is said to be **adaptive**. The term *adaptive* is also used to describe the result of learning (see Chapter 6), but here we are concerned with the evolutionary sense of the word. By definition, natural selection will favour individuals that are well adapted to the environment in which they live, so that, if the adaptive characteristics they show have an inherited component, they will tend to spread through the population. The environment is not static, however—it changes with time as a result not only of physical factors such as weather, but also of the activities of the animals and plants themselves. So the adaptiveness of a particular behaviour will depend upon the animal's immediate environment. The same behaviour in different environmental circumstances may result in different costs and benefits, and so may be more or less adaptive.

Each of the various phenotypic characteristics that make up an individual can affect its fitness. You can think of each characteristic fulfilling one or more *functions* in terms of the immediate benefits resulting from it which tend to improve fitness. One function of incubation behaviour in birds, for example, is to keep the eggs at the correct temperature for development. Another may be to protect eggs from predators. Each characteristic also has fitness costs associated with it, but the term function only refers to the fitness benefits. Incubation behaviour, for example, reduces the time that the bird has available for feeding, which may have a variety of effects; in the short-term the bird may lose weight so that, in the long-term, it is less likely to be able to breed again. Losing weight is a *consequence* of incubation, it is not its *function*.

It might be thought that, since natural selection will be continually operating against less fit individuals and in favour of those best adapted to their environment, a population should eventually evolve into one consisting entirely of identical individuals, all equally fit. This does happen to some extent—for example all blackbirds now incubate their eggs. But there will still be variation—

for example blackbirds vary in how *effectively* they incubate. There are many reasons for this variation, for example:

1 Differences in fitness between individuals will not always be due to inherited characteristics.

2 Genetic drift may operate.

3 The environment constantly changes and selection may fluctuate, affecting different members of a population at different times.

4 Individuals may achieve equal fitness in a particular environment by different means, i.e. the population may consist of a number of equally fit polymorphs. This last point will be discussed further in Section 4.4.4.

4.4.1 Assessing the adaptiveness of behaviour: studying natural variation

One way of assessing the adaptiveness of a particular behaviour pattern is to look at the natural variation within a species in relation to variation in environmental conditions.

For example, P. K. Ducey and E. D. Brodie studied the behaviour of the two-lined salamander (*Eurycea bislineata*) in response to one of its predators, the garter snake (*Thamnophis ordinoides*) (Ducey and Brodie, 1983). Salamanders respond to a garter snake in two main ways—they either run or they remain immobile. Table 4.1 shows that the effectiveness of each type of behaviour depends on the kind of contact the salamander has with the garter snake.

Table 4.1 The percentage survival of salamanders showing each type of behaviour in response to two types of contact with a garter snake. A touch from the head or body of a garter snake merely indicates that a predator is near by, but a tongue-flick indicates that the garter snake has detected the salamander.

Type of contact with garter snake	Salamander runs	Salamander remains immobile
Flicked with tongue	90% survive	0% survive
Touched by head or body	34% survive	100% survive

☐ From the results in Table 4.1, what do you deduce is the most adaptive behaviour for a salamander to show in response to (a) being flicked by a garter snake, and (b) being touched by the head or body of a garter snake?

■ (a) A salamander should run if it is flicked by the tongue of a garter snake because it then stands a 90% chance of surviving instead of a 0% chance if it remains immobile. (b) It should remain immobile if it is touched by the head or body of a garter snake because it then stands a 100% chance of surviving instead of only a 34% chance of surviving if it runs.

T. G. Dowdey and E. D. Brodie then looked at individual variation in the response to garter snakes among salamanders brought into the laboratory from two different areas of New York State in the USA, Hagaman and Shokan (Dowdey and Brodie, 1989). Individual salamanders that responded to a tongue-flick from a garter snake (hand-held to prevent the snake actually killing the salamander) by running always responded to a tongue-flick in the same way, but the two populations showed different proportions of 'runners' versus 'stayers'. Individuals from Hagaman were more likely than those from Shokan to run in response to a tongue-flick and, as a result, only 40% of Shokan salamanders would be likely to survive contact with a free-moving garter snake, whereas 71% of Hagaman salamanders would. Dowdey and Brodie assessed the density of garter snakes in the two locations and found that snakes were common in Hagaman but were never found in Shokan. The response to tongue-flicks of running may have a genetic basis and may have been selected for in the Hagaman population as a result of heavy snake predation, or it may simply have been acquired by individual salamanders as a result of experience. Appropriate experiments to distinguish between these two possibilities have yet to be carried out.

4.4.2   Assessing the adaptiveness of behaviour: the comparative approach

Another way of assessing the adaptiveness of a particular behaviour pattern is by using a *comparative* approach. This can be done in two ways:

1   Comparing the species in question with a closely related species that occupies a different habitat. The term *habitat* describes the type of environment in which an organism lives. If the related species does not show the particular behaviour, this suggests it is specific to the habitat of the species under consideration, and is related to the different ways in which natural selection has affected the two species.

2   Looking at other, closely related, species that have been faced with similar selection pressures to see whether they too evolved the same behaviour.

An example of the first type of comparative approach is the study by Esther Cullen (1957) of the nesting behaviour of the cliff-nesting kittiwake (*Rissa tridactyla*), in the Farne Islands off the coast of Northumberland, which she compared with that of various other gulls which nest on the ground, such as the black-headed gull (*Larus ridibundus*) and the herring gull (*Larus argentatus*).

The kittiwakes in Cullen's study population nest on very narrow cliff ledges (Figure 4.12) and she observed that their eggs are not preyed upon by ground predators such as rats or aerial predators such as herring gulls, both of which frequently take the eggs of ground-nesting birds. Nor do the kittiwakes, which feed mainly on fish and plankton, cannibalize the eggs and young chicks of their neighbours, unlike many ground-nesting gulls. The kittiwake shows few of the behaviours that appear to be adaptive in its ground-nesting relatives—it does not camouflage its nest, for example, it defaecates on the ledge around the nest, and it rarely gives alarm calls.

☐   Why might it be adaptive for a ground-nesting bird to defaecate only at a distance from its nest?

■   Because white bird droppings, being not only smelly but visually very conspicuous, may attract predators to the nest.

Figure 4.12   Kittiwakes nesting on cliff ledges.

On the other hand, kittiwakes do show a number of behaviour patterns that appear to adapt them to cliff-nesting. Their fighting behaviour is limited and very stereotyped compared with ground-nesting gulls—more suited to the confined space of a small ledge. They build cup-shaped nests of grass, glued together with mud, whereas their ground-nesting relatives, which do not need to prevent eggs falling off a cliff ledge, have simple nests consisting of a simple scrape in the ground or a small pile of grass or seaweed. Young kittiwakes stay in the nest for longer than ground-nesting gulls and eat regurgitated food directly from the parent's throat instead of from the ground where it is dropped by ground-nesting parents. They also stay in the nest when danger threatens, unlike the young of ground-nesting gulls, which run away from the nest and hide.

☐   What advantages would the behaviour shown by young kittiwakes have to a cliff-nesting bird?

■   Young ground-nesting birds can leave the nest before they are able to fly, unlike kittiwakes which would be in danger of falling off the ledge. Parents attempting to drop food onto a narrow ledge are likely to lose it, and young kittiwakes have nowhere to hide other than the nest.

Evidence provided by the first type of comparative approach can be strengthened if evidence from the second type of approach is also available. For example, Jack Hailman (1965) studied the swallow-tailed gull (*Larus furcatus*) of the Galapagos

Islands. This relative of the kittiwake also nests on cliffs, but they are not so steep or so high above the ground as the cliffs on which the kittiwake nests. As a result, the swallow-tailed gull is subject to greater predation pressure than the kittiwake, though not so great as that affecting ground-nesting gulls. Hailman therefore predicted that swallow-tailed gulls would show behaviour relevant to avoiding predation intermediate between those of kittiwakes and those of ground-nesting gulls. He found that swallow-tailed gull chicks defaecate over the edge of the ledge, whereas kittiwake chicks defaecate on the nesting ledge, making it very conspicuous, and ground-nesting gulls defaecate well away from the nest. They also eat food from the parent's bill, whereas kittiwakes eat food direct from the parent's throat and ground-nesting gulls eat it from the ground.

☐   Do these observations support Hailman's prediction?

■   Yes. Both defaecation and feeding behaviour are intermediate between those of kittiwakes and ground-nesting gulls.

Another example of the comparative approach comes from John Crook's (1964) work on 90 species of weaverbirds in Africa and Asia. Different species of this small bird are all fairly similar in appearance but show considerable variation in social behaviour. Crook found that species from open grassland (savannah) tend to eat seeds, live in flocks and be polygynous, with marked differences in appearance between males and females; they breed in large conspicuous colonies. Species living in the forest tend to eat insects, be solitary and monogamous (one male mates exclusively with one female, at least over the period of one breeding cycle—see Chapter 9), with males and females similar in appearance; they have camouflaged nests in large territories defended from other pairs.

Crook argued that insect food in the forest is widely but fairly evenly spread out and so both parents are required to carry food to the nest over the relatively long distances necessary if the young are to be provided with enough food. To make sure there will be an adequate supply of insects, the parents also have to defend a large territory. To provide maximum protection against predators, the nest is camouflaged and the adult birds are dull-coloured and do not attract attention when they visit the nest. In the savannah, on the other hand, seeds are dispersed in patches—abundant at some times and at some places, absent at others. Under such conditions birds can obtain food most efficiently by forming flocks—many pairs of eyes are better than one when looking for seeds and, when seeds *are* found, there are enough available to feed the whole flock.

Since nesting sites protected from predators, i.e. trees, are scarce in the savannah, many birds nest in the same tree. Because food is abundant, a female can feed the young by herself, leaving the male free to go off and court other females. Typically males compete for nest sites within a colony, and the successful males perform an elaborate display next to the nest to attract females.

The comparative method is a very useful way of gaining insight into the adaptiveness and function of behaviour, but it does have its limits. First, it is often difficult to disentangle cause and effect. For example, a correlation between diet and size of feeding group like that observed in the weaverbirds could be interpreted in two different ways. Group size could have evolved in response to diet, e.g. for a seed-eating bird, seeds can be exploited more efficiently by flocks. Or, diet could have evolved from group size, e.g. birds feed in groups because it is better for avoiding predators, but then need to eat the sort of food that is sufficiently common to allow them to feed together. Secondly, the differences between species observed *now* are the results of events that occurred a long time ago and there is no way of measuring selection pressures operating in the past. And lastly, just because differences in behaviour and ecology are correlated between two related species, it does not *necessarily* mean that the characters under study reflect adaptation to different environmental conditions experienced by the two species during their evolution, though it does suggest it. Ideally, comparison of two species should lead to predictions about a third species that has not yet been studied—like Hailman's study of the swallow-tailed gull—so that any hypothesis arising from the comparative study can be tested independently on the third species.

4.4.3 Assessing the adaptiveness of behaviour: the experimental approach

Further insight can be gained into the adaptiveness of behaviour by using an experimental approach, especially if it is combined with other approaches. The selection pressures that actually operated during evolution cannot be measured, but experiments can be carried out to investigate present-day selection pressures. A classic series of experiments was carried out by Niko Tinbergen and his colleagues (1962) to study eggshell removal behaviour in the black-headed gull (*Larus ridibundus*) (Figure 4.13).

Figure 4.13   Black-headed gull removing an eggshell from its nest.

Many birds remove empty eggshells after the chicks have hatched, usually by picking them up in their beaks, then flying some distance away from the nest before dropping them, even though this leaves their chicks unguarded for a few minutes. This suggests that there must be considerable advantage in removing eggshells, which outweighs the risk of exposing the chicks to possible predation during the parent's absence. Tinbergen and his colleagues thought of several possible reasons why eggshell removal might be beneficial. For example the sharp edges of the shell could injure the chicks; an unhatched egg could fall into an empty shell, giving it a 'double' shell and making it difficult for the chick to break out; the shells might get in the way when the parent is trying to brood the young; they might be a breeding ground for disease; or the conspicuous white inside of the shell, in contrast to the camouflaged exterior, might attract the attention of predators.

The researchers knew from previous observations that the gulls removed not only eggshells but also many other 'foreign' objects of similar size, and that the behaviour is shown for several weeks both before and after hatching.

☐ Which of the possible reasons for eggshell removal does the latter make less likely?

■ Injury or disease in newly-hatched chicks, prevention of hatching or interference with brooding, since the behaviour is performed several weeks *before* the chicks hatch.

Further clues about the functions of eggshell removal behaviour could also be obtained using the comparative approach.

☐ Remember that kittiwakes are not subject to severe nest predation. Would you expect them to remove eggshells from near the nest if this behaviour functions (a) to keep the nest camouflaged against predators, or (b) to prevent disease or injury to the young?

■ (a) If the only function of eggshell removal is to keep the nest camouflaged, kittiwakes should not remove eggshells because there is no need. (b) If one of the functions is to prevent disease or injury, kittiwakes should remove them.

In fact, kittiwakes do *not* remove eggshells, even though they keep their nests very clean and do throw away any 'foreign' objects. This provides support for the hypothesis that eggshell removal in black-headed gulls functions to maintain camouflage of the nest.

Tinbergen and his colleagues set out to test this hypothesis experimentally. They laid out black-headed gull's eggs over an area of land next to the gull colony. In one experiment, half the eggs were left in their natural state and half were painted white (and so were very conspicuous); in another experiment, all the eggs were natural and slightly concealed by vegetation but half of them had a broken eggshell placed nearby. The eggs were watched from a hide to see which ones were taken by predators. The results are shown in Table 4.2.

Table 4.2 Results of Tinbergen's experiments on the effects of camouflage and the presence of broken eggshells on predation of black-headed gull's eggs.

Natural eggs		White eggs		With eggshell		Without eggshell	
Taken by predators	Not taken	Taken by predators	Not taken	Taken by predators	Not taken	Taken by predators	Not taken
13	55	43	26	39	21	13	47

☐ Study Table 4.2. Which eggs were more vulnerable to predation: (a) White eggs compared with natural eggs? (b) Eggs with a broken eggshell nearby compared with eggs without?

■ (a) White eggs. (b) Eggs with a broken eggshell nearby.

In another experiment, Tinbergen and his colleagues were also able to show that carrion crows (*Corvus corone*), one of the main predators on gull eggs, quickly learn to associate broken eggshells with unhatched eggs. First, they put out eggshells alone, which the crows ignored. Then they put out eggshells paired with an unhatched egg, some of which were discovered and eaten by crows. Then, eggshells were again put out alone. This time the crows paid particular attention to them, searching the area nearby, suggesting that the crows had formed a search image for the eggshells (Section 2.7.2). Taken together, these experiments provide evidence that the presence of an eggshell near the nest of a black-headed gull makes the nest more likely to be predated because it makes it easier to spot, especially from the air.

4.4.4 Adaptive strategies

There are usually many ways in nature of achieving the same end. To achieve the maximum lifetime reproductive success possible, for example, some animals have evolved so that a very large proportion of the energy available for reproduction is put into producing as many offspring as possible, and none into parental care of those offspring. Usually most of the offspring die but the chance is that, out of such a large number, some will survive. Many fish and insects breed in this way. Other animals have evolved to produce only a few offspring and to put most of the energy available for reproduction into parental care. The chances are very much greater that each offspring will survive, but only a few offspring can be produced in this way. This is the way that most mammals and birds breed. Each species is said to adopt a particular reproductive *strategy*. Use of the word 'strategy' here simply refers to the collection of phenotypic characteristics involved in reproduction, and does not necessarily imply that alternatives are possible within a single species. During evolution, natural selection has led to whichever mode of reproduction is most adaptive becoming fixed in each species. Nor does the term imply any conscious decision-making process on the part of members of a species.

Similarly, individuals within a species may not all behave in exactly the same way. They may adopt different *behavioural* strategies, which is just another way of saying they have different behavioural phenotypes. Again, the term does not imply making any conscious decision or adopting any kind of plan—species are just polymorphic for a particular behaviour pattern or set of behaviour patterns. For example, you will remember from Section 4.2.1 that male ruffs are either territorial or satellite.

☐ What might cause males to adopt satellite behaviour?

■ Males might be satellites because they cannot get territories of their own.

Thus it is possible that satellite males cannot obtain territories because they are young, inexperienced males, or because they have lost out in the competition for territories. They only adopt satellite behaviour as a 'second best' strategy, while they wait for the opportunity to get a territory of their own. This seems to be the case in another lekking bird that has territorial and satellite males, the black grouse (*Lyrurus tetrix*).

☐   Why is this possibility unlikely for ruffs?

■   Satellite males have different plumage from territorial males and never develop into territorial males.

There is a second possible reason for adopting satellite behaviour. Perhaps, by being satellites, they gain just as many mates as if they were territorial. In other words, they could be achieving reproductive success equal to that of territorial males by using an alternative means of obtaining mates. This is probably true for the ruff.

The adaptiveness of a particular behaviour pattern is not necessarily fixed. As mentioned before (Sections 4.3.4 and 4.4), it is dependent on the environmental circumstances at the time. The adaptiveness of a particular behavioural strategy is also dependent on the behavioural strategies shown by other members of the population. For example, when there are only a few territorial male ruffs relative to the number of satellites, territorial males probably have higher reproductive success than satellites because, being undistracted by territorial fights, they are better able to monopolize females. The proportion of territorial males therefore increases. On the other hand, when satellite males are rare, their reproductive success is probably higher than that of territorial males because the territorial males spend a lot of time and energy fighting over females, providing plenty of opportunity for satellites to mate. So the proportion of satellites increases. In this way a balance between territorial males and satellites is maintained.

Summary of Chapter 4

There is variation within species both between different geographical areas and between individuals within a given area. Some of this variation is due to differences in genotype and some to differences in environment. In a population consisting of a variety of phenotypes, some individuals will be better able to compete than others. These individuals will leave more offspring than others, i.e. they will have greater reproductive success, and, if their competitive ability is inherited, their phenotype will increase in the population as a result of natural selection. The relative fitness of an individual is equal to its lifetime reproductive success as a proportion of the lifetime reproductive success of the most successful individual in the population. Relative fitness depends not only on the individual's own phenotype but on the proportions of other phenotypes in the population. The same fitness may be achieved by different means.

Natural selection may be stabilizing, directional or disruptive. Evolution (i.e., phenotypic change over time) only occurs in the last two cases. Altruistic behaviour can increase an individual's inclusive fitness if relatives are helped by the behaviour. Such kin selection can explain the evolution of altruistic behaviour such as alarm calling. Group selection cannot explain the evolution of altruistic behaviour.

Behaviour is adaptive if the benefits of showing it, in terms of increased fitness, outweigh the costs. The adaptiveness of behaviour can be assessed by looking at variation within a species in relation to differences in habitat; comparing closely related species that occupy different habitats and relating differences in behaviour to differences in habitat; comparing closely related species in similar habitats that are presumed to have been faced with similar selection pressures to see if they evolve similar behaviour; and by carrying out experiments.

Objectives for Chapter 4

When you have completed this chapter, you should be able to:

4.1   Define and use, or recognize definitions and applications of each of the terms printed in **bold** in the text.

4.2   Describe and give examples of the types of variation found within a single species. (*Question 4.1*)

4.3   Explain how the process of natural selection operates and how it relates to the process of evolution. (*Question 4.2*)

4.4   Recognize and give examples of components of fitness. (*Question 4.3*)

4.5   Calculate relative fitness. (*Question 4.4*)

4.6   Calculate a selection coefficient. (*Question 4.4*)

4.7   Recognize and distinguish between different types of natural selection. (*Question 4.5*)

4.8   Distinguish between natural selection operating at the individual level and group selection, and explain why group selection is unlikely to have been important in the evolution of altruism. (*Question 4.2*)

4.9   Give an example of altruistic behaviour, analyse such behaviour in terms of its costs and benefits, and explain how it may have evolved. (*Question 4.6*)

4.10   Calculate the coefficient of relatedness, *r*, between two related individuals. (*Question 4.7*)

4.11   Give examples of studies where lifetime reproductive success and survival value have been measured.

4.12   Analyse the adaptiveness of a behaviour pattern in terms of its fitness costs and benefits. (*Question 4.8*)

4.13   Describe ways in which the adaptiveness of behaviour can be assessed and give examples in each case. (*Question 4.9*)

4.14   Give examples where different individuals adopt different adaptive strategies and analyse the adaptiveness of each strategy in terms of its fitness costs and benefits.

Questions for Chapter 4

Question 4.1 (*Objective 4.2*)
Are the following examples of (a) polytypy; (b) polymorphism; or (c) neither?

1   Some individual two-lined salamanders always run from a garter snake while others always remain immobile.

2   Banded killifish nearly always flee from a predator while three-spined sticklebacks nearly always remain immobile.

3   Kittiwakes do not camouflage their nests while black-headed gulls do.

4   Great tit (*Parus major*) males in one wood sing six different song types while great tit males in a wood 200 kilometres away sing only one song type.

5   Black grouse males are either territorial or satellite during the breeding season.

Question 4.2 (*Objectives 4.3 and 4.8*)
Which of the following statements are true and which are false?

(a) If predators take only the slowest-running individuals, the average running speed of the population may increase over time.

(b) The purpose of natural selection is to improve the species.

(c) In a constant environment there is no evolution because there is no natural selection.

(d) Where there is natural selection there will always be evolution.

(e) Evolution can occur without natural selection.

(f) Natural selection is just the process of 'survival of the fittest'.

(g) Natural selection will always favour altruism towards other group members if, as a result, the group as a whole increases its reproductive rate.

Question 4.3 (*Objective 4.4*)
(a) What are the two major components of fitness?

(b) Which of the following are components of fitness? (i) Type of parental care; (ii) number of females mated with; (iii) ability to obtain a mating territory.

Question 4.4 (*Objectives 4.5 and 4.6*)
In a population of 10 pairs of great tits, pairs 1–10 produced 2, 6, 1, 0, 3, 0, 4, 0, 1, and 3 surviving offspring respectively. What is the relative fitness of and the selection coefficient for (a) pair 2; (b) pair 10; (c) a pair producing the average number of surviving offspring?

Question 4.5 (*Objective 4.7*)

Males of a certain species of bird repeatedly raise their tail feathers when displaying to females. Table 4.3 shows the changes in a population, over a period of 200 years, in the numbers of males that raise their tail feathers at various rates. What type of selection, if any, is operating on the population (a) over the first 100 years; (b) over the second 100 years?

Table 4.3 (For use with Question 4.5)

Tail-raising rate	Number of males		
	Year 0	Year 100	Year 200
Very slow (1–4 per min)	25	15	20
Slow (5–8 per min)	100	60	75
Medium (9–12 per min)	100	100	60
Fast (13–16 per min)	25	60	75
Very fast (17–20 per min)	0	15	20

Question 4.6 (*Objective 4.9*)

In some bird species, young adults hatched the previous year do not breed themselves, even though they are sexually mature. Instead they help breeding birds to feed their offspring. What is the main fitness cost to the young adults of this behaviour and under what conditions might they benefit?

Question 4.7 (*Objective 4.10*)

What is the coefficient of genetic relatedness, r, between aunt and niece, where the niece is the daughter of a half sibling?

Question 4.8 (*Objective 4.12*)

List the probable costs and benefits of the following behaviour patterns and assess whether they are likely to be adaptive: (a) a territorial ruff fights with another territorial male; (b) a female red deer fails to conceive one year and, instead of weaning her offspring of the previous year, carries on suckling it for several months.

Question 4.9 (*Objective 4.13*)

What four means of assessing the adaptiveness of behaviour were described in this chapter?

References

Clutton-Brock, T. H., Guinness, F. E. and Albon, S. D. (1982) *Red deer. Behaviour and ecology of two sexes*, The University of Chicago Press.

Crook, J. H. (1964) The evolution of social organization and visual communication in the weaverbirds (Ploceidae), *Behaviour Supplement*, **10**, pp. 1–178.

Cullen, E. (1957) Adaptations in the kittiwake to cliff-nesting, *Ibis*, **99**, pp. 275–302.

Dowdey, T. G. and Brodie, E. D. (1989) Antipredator strategies of salamanders: individual and geographical variation in responses of *Eurycea bislineata* to snakes, *Animal Behaviour*, **38**, pp. 707–711.

Ducey, P. K. and Brodie, E. D. (1983) Salamanders respond selectively to contacts with snakes: survival advantage of alternative antipredator strategies, *Copeia*, **1983**, pp. 1036–1041.

Hailman, J. P. (1965) Cliff-nesting adaptations of the Galapagos swallow-tailed gull, *Wilson Bulletin*, **77**, pp. 346–362.

McLean, E. B. and Godin, J-G. J. (1989) Distance to cover and fleeing from predators in fish with different amounts of defensive armour, *Oikos*, **55**, pp. 281–290.

Sherman, P. W. (1977) Nepotism and the evolution of alarm calls, *Science*, **197**, pp. 1246–1253.

Tinbergen, N., Broekhuysen, G. J., Feekes, F., Houghton, J. C. W., Kruuk, H. and Szulc, E. (1962) Egg shell removal by the black-headed gull *Larus ridibundus* L.; a behaviour component of camouflage, *Behaviour*, **19**, pp. 74–117.

Further reading

Dawkins, R. (1986) *The Blind Watchmaker*, Penguin Books.

Maynard Smith, J. (1993) *The Theory of Evolution*, Canto, Cambridge University Press (but also in several earlier editions by Penguin Books).

Patterson, C. (1978) *Evolution*, Routledge and Kegan Paul in association with the British Museum (Natural History).

Ridley, M. (1992) *Evolution*, Blackwell Science Inc.

CHAPTER 5
DEVELOPMENT OF BEHAVIOUR

5.1   Different species, different lives

In this chapter, we shall look at the changes that occur in an organism during its life—its **development**—and in particular at its behavioural development. In the previous chapters, examples were given from different species of animals with no reference to any differences in life cycle between the species mentioned. When considering the development of behaviour, however, it is important at least to be aware that different animals are born at different states of maturity. For example the 2–cm long newborn kangaroo that paws its way up from the mother's vagina to her pouch and clasps its jaws to a nipple is at a different stage of development from a virtually helpless yet fully formed newborn kitten, which is itself at a different stage of development from a highly mobile newborn foal. Similarly some birds are **precocial** and undergo most of their development inside the egg so that they are mobile immediately upon hatching (e.g. quail). Other birds are **altricial** and hatch as immature, uncoordinated fledglings (e.g. blackbird). In arthropods, some species hatch as the adult form (e.g. spiders) whereas other species such as butterflies and moths have a number of intermediate forms, e.g. larvae (caterpillars) and pupae, before the adult form is reached.

One consequence of these highly varied life cycles is that comparisons between species need to take account of the fact that a given developmental milestone (e.g. birth) may not occur at the same developmental stage in different species. A second consequence is that some stages of development might be more responsive to environmental stimuli than others. As the various stages occur at different times in different organisms it follows that different species are going to respond to environmental stimuli at different times. Any one species could be sensitive to a number of different environmental factors at any of a number of stages in its development. But a different species might be sensitive to different factors at different times.

5.2   Developmental processes

The fertilized egg, the zygote, is poised to follow any of a number of developmental routes to adulthood. The *exact* route is not fixed at the outset because it is a product of the continuous interactions between the genotype and the constantly changing environment of the developing organism. The influence of the genotype can be seen in the form of the adult, whether it be a particular type or colour of plumage (e.g. the unmistakeable plumage of the peacock), or a behaviour pattern typical of a particular species (e.g. the nest-building of a weaverbird, or of a termite), but that does not mean that all peacocks are identical or that all termite nests are the same. The effects of the genotype are influenced by the environment in which it operates. Nor must the environment be thought of as a constant arena in which the genotype functions, for the environment changes moment by moment. The instant a gene starts to produce a new protein the environment of the gene has changed because the environment now contains the new protein. The moment a bird inserts a new twig into its nest, the environment of the bird has changed—the nest is different. Development, then, does not follow a blueprint or a programme in some inevitable way; if it did how could you account for phenocopies (Section 3.6.4)? No, development is a result of the continuous transaction between the genotype and the environment.

C.H. Waddington from Edinburgh likened the developmental process to a ball rolling down a series of valleys and gullies (Figure 5.1). The ball is the organism. This analogy captures the inexorable nature of the developmental process (the ball has to keep rolling down the valley, the organism has to keep on developing) and also the fact that at each point on the journey the environment is different from each other point. Another feature the analogy has in common with the real process is that at choice points the ball takes only one of several possible valleys and cannot reverse the process and take another choice, just as developing cells differentiate (change from generalized to specialized cells) irreversibly. Where the analogy breaks down is in considering the organism as a passive participant, as a passenger being carried along by the tide of events; the reality is that the organism is a major player in the process of development.

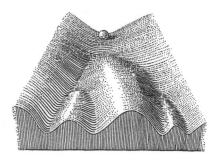

Figure 5.1 Waddington's model of the developmental process.

There are essentially four types of changes occurring in an organism during development and the important point is that these changes happen simultaneously.

1 **Growth**. The organism is increasing in size as the result both of an increase in the number of cells (cell division is occurring) and an increase in the size of those cells.

2 **Differentiation**. The zygote has no nerve cells, no liver cells, no blood, no fur. At some point some of the cells produced by the zygote by cell division must change their form or the way they function to become different sorts of cell. They do this by changing which proteins they produce: liver cells produce a different combination of proteins from nerve cells, for instance. The change is gradual, requiring many intermediary cell types, and many cell divisions, eventually giving rise to the final, differentiated cell types. This whole process is called differentiation.

3 **Maturation**. Maturation is the combined effect of growth and differentiation. For instance, cells may have grown and differentiated to form a muscle, but unless that muscle has a blood supply and is connected to the nervous system it is not mature. The formation of connections between nerve cells and between nerves and muscles is also a maturational process.

4 Experience. This refers to any external influence that leads to learning, i.e. a change in the nervous system that results from its interaction with the environment (which will be discussed in Chapter 6). The connections between nerves and muscles are often only firmly established during development as a consequence of their functioning; experience will therefore shape the way the nervous system develops (this will be discussed further in Book 4). At the level of the individual, learning refers to changes in behaviour that result from an animal's interaction with its environment.

The first two processes will not be considered further here, though they do recur in subsequent books. In this section we are concerned with maturation and experience.

5.2.1   Maturation

Maturation itself is responsible for some changes in behaviour that might otherwise be thought of as the result of experience. Thus any consideration of the causes of changes in behaviour during development must take account of the possibility of maturation.

For instance the eyes of frog tadpoles are unresponsive to very small objects moved across their field of view. When those tadpoles become frogs, however, their eyes are very responsive to those same small objects. Tadpoles graze on algae that grow on surfaces and small objects have little relevance to them, whereas frogs feed on small moving objects, such as flies, and responding to them has obvious advantages. Thus, both as tadpoles and as frogs the responsiveness of the eyes is adaptive. The change in responsiveness is the result of maturation.

Many young birds can be seen flapping their wings before they can fly. It would be easy to assume that this behaviour is essential to strengthen key muscles and improve coordination. Although this may be true for some species, it certainly is not true for the pigeon (*Columba livia*), as revealed by an experiment in which two groups of birds were reared to fledging. The first group had normal experience and the opportunity to practice flapping. The second group were reared in tubes so they could not move their wings, and thus had no opportunity to practice flapping. When the first group were flying satisfactorily, the second group were released from their tubes and were able to fly immediately. Thus actual flapping is not a necessary prerequisite of flight.

Young chickens (*Gallus gallus domesticus*) are precocial and fend for themselves soon after birth. Many parent birds feed their young, but the hen does not feed her chicks. The chicks can survive for several days without ingesting any food, but clearly the sooner they start feeding the better. At first the pecks of chicks are fairly inaccurate, but their aim rapidly improves. The question is, do they require experience of pecking for the improvement to occur?

By simply keeping chicks in the dark from the time of hatching to stop them pecking, Cruze (1935) was able to separate the experiential effects of pecking from maturational changes. His results are presented in Figure 5.2.

☐   Study Figure 5.2 and decide whether it provides evidence for a maturational or an experiential improvement in pecking.

■   It provides evidence for both. Older chicks were more accurate peckers than younger chicks, whether they had had experience of pecking in the light or not. However, chicks of the same age who had experience of pecking in the light were more accurate peckers than those without that experience.

It has been argued by some scientists that practice at pecking begins inside the egg. Active head movements can be seen in the egg, along with opening and closing of the beak, movements similar to pecking. It is certainly true that some coordinated movements are possible in the egg long before they are needed, as discovered for chicks. Recordings from chick embryo leg muscles using fine wire electrodes suggest that coordinated hatching movements are present 12 days before hatching. These findings are important because they remind us that development is a continuous process, with behaviour occurring long before hatching or birth.

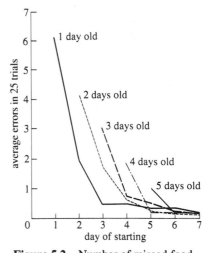

Figure 5.2   Number of missed food pecks by chicks of different ages before and after practice. Two-day-old chicks had been kept in the dark for 1 day and only allowed to peck on Day 2, 3-day-old chicks had been kept in the dark for 2 days and allowed to peck on Day 3, and so on.

One way to eliminate all practice effects is to anaesthetize the animal. Tiger salamanders (*Ambystoma tigrinum*) have been raised from eggs to swimming larvae in water containing the anaesthetic chloretone. These experimental larvae were unable to go through the normal rudimentary swimming movements as they developed. They were transferred to plain water at an age when swimming is usually performed as a complete movement. As soon as they recovered from the anaesthetic, they swam normally.

This example illustrates that some changes or improvements in behaviour occur in the developing organism simply as a result of maturation. The next section considers the role of experience in the development of behaviour.

5.2.2 Experience

The behaviour of an organism can be affected by the amount of practice it has had, by stimuli that have impinged on it and by its opportunities for learning, just as much as it can be affected by its stage of development. Sometimes it is quite easy to overlook the kind of experience an organism has had, by ignoring embryonic movements or non-obvious stimuli for instance. Because much of the rest of this chapter deals with experiential factors, only two, very different examples are given here.

The A/J strain of mice (*Mus musculus*) exhibits a low level of aggression (as measured by the frequency with which individuals chase, attack and fight with other mice, a sequence called chase–attack–fight) when previously isolated males are put into groups of four. In contrast the CFW strain exhibits high levels of aggression under the same circumstances. When an A/J female is mated to a CFW male, the male offspring have a low aggression score. When an A/J male is mated to a CFW female, the male offspring have a high aggression score.

☐ Is the aggression score of the male offspring more similar to the strain of the mother or the father?

■ The aggression score of the offspring is more similar to the aggression score of the strain of the mother. If the mother is of the low aggression score strain, then the offspring exhibit a low aggression score. If the mother is of the high aggression score strain then the offspring have a high aggression score.

A genetic influence on aggression passed from mother to offspring could not account for this result.

☐ Suggest an alternative explanation for the result.

■ Something to do with the mother's pregnancy, or how she treats her offspring after the pups are born, influences the later aggressiveness of the offspring.

It has also been shown that A/J male mice suckled and raised from birth by CFW females (referred to as a *cross-fostered* group) had a significantly higher aggression score (an average of 27 chase–attack–fights per hour) than A/J males suckled and raised by A/J females that were not their mothers (an *in-fostered* group), which averaged 18 chase–attack–fight sequences per hour. Thus the strain of foster-mother the pups had from birth, rather than the genotype of their real mother, influenced their subsequent aggressiveness.

A second example of how experience can affect behaviour concerns locusts (*Locusta migratoria*). Most generations of this species of locust are solitary, living in grassland with a population scattered over a large area. Occasionally, at intervals of between five and eight years, gregarious locusts arise which congregate into huge swarms (one such was recorded as 90 km long and 5 km wide). The difference is not just in behaviour; at one time the solitary and swarming locusts were considered to be two separate species because they differed in appearance. So why do some generations of locusts develop into the solitary form while other generations develop into the gregarious form? A crucial factor seems to be the amount of contact a young locust has with conspecifics. Frequent encounters with other locusts while young alters development and results in gregarious adults. In their normal, dry, sparse environment, the survival of young locusts is not high, so they rarely encounter other young locusts and they develop into the solitary form. If conditions change, for instance after good rains, the survival rate is much higher and more locusts encounter other locusts when young. As a consequence they develop into the gregarious form.

5.2.3   A developmental timetable

There is a sequence of developmental events which is followed by each organism as it develops. This is true at the cellular level as different cells go through different stages of differentiation, at the morphological level (e.g. as the down of a chick is replaced by feathers), as well as at the behavioural level. Thus babies suck before they can smile, sit up before they can walk; kittens can move around before they can see; ducklings can swim before they can fly. The timetable for development for each species appears to be fairly stable, with particular events following particular other events. However, this concept of a relatively stable developmental sequence needs two important qualifications.

The first qualification is that the developmental sequence is not progressing along a fixed path towards a prescribed end-point (i.e. the developmental sequence is not determined). There are many ways in which the sequence could be diverted to follow a different path (as in phenocopies) but, given a reasonably stable genotype, generation by generation, and a reasonably stable environment, generation by generation, there is little variation in the sequence.

The second qualification is that the timing of the events is not fixed. Exactly when a particular developmental event will occur cannot be predicted. One need only consider the first occurrence of walking in human children! The sequence of events may be stable but the timing of individual events is not. This does not mean that individual events cannot be used as an indication of development and maturity. Such events (e.g. the grasping reflex of a baby, or the Babinski reflex, in which, if the sole of the foot is touched, an infant will extend its big toe and fan the other toes) are regularly used by paediatricians as indicators of maturity. But it does mean they have to exercise caution when interpreting these reflex tests.

These two qualifications are well illustrated by studies on the effects of undernutrition in young rats (*Rattus norvegicus*). If rats are given insufficient food from birth to weaning (a period of about 25 days) they grow poorly and remain stunted for the rest of their lives, irrespective of how much they subsequently eat. The sequence of developmental events though, both morphological (e.g. ear unfolding, tooth eruption, eye opening) and behavioural (e.g. grasping objects and showing a startle response to a sudden noise) is unaffected. However, these events occur later in the undernourished animals than in adequately fed animals, with the result that undernourished animals take longer to mature.

5.3   The environment and the developing organism

You should remember from Chapter 4 that the environment is everything that surrounds an animal, including the physical world and all other organisms and the individual animal's internal environment.

You should also recall from Chapter 2 that stimuli that emanate from the external environment (external stimuli) are detected by exteroceptors, and that stimuli that emanate from the internal environment (internal stimuli) are detected by enteroceptors and proprioceptors. Both kinds of stimuli have an influence on development, though not all stimuli influence or can be detected by all animals.

5.3.1   Internal stimuli

Internal stimuli operate both locally and at a distance from their source during development. At a cellular level, there are signals within cells that determine which genes should be active or inactive and which enzymes should be working. On a larger scale, there are signals which operate between cells that affect whether cell division or differentiation should occur, and where cells should migrate, or, in the case of nerve cells, where they should make connections. And, on a larger scale still, there are hormones circulating in the blood that determine, for example, the rate of uptake of glucose by cells or the rate of production of urine.

The number and variety of internal stimuli is vast, for, in a sense every chemical within an organism is a stimulus for some chemical reaction or other within the organism. Hormones are obvious internal stimuli, but proteins, vitamins, sugars, minerals and even water act as stimuli for some reactions. For the most part, the reactions keep the stimuli within certain limits and ensure that the internal environment is more or less stable, a process which is known as *homeostasis* (this will be discussed further in Chapter 7).

Departures from normal levels of any substance within the body can profoundly affect development. For example, the amino acid phenylalanine, although essential to life in humans, is normally maintained at a low level where it acts as a necessary link in a number of biochemical pathways and as a constituent of proteins. But if the enzyme phenylalanine hydroxylase, which converts phenylalanine to tyrosine (Section 3.4 and Figure 3.11) is absent, an excess of phenylanaline builds up and this causes abnormal human development. Any developing organism will die if some substances are not maintained within strict limits: oxygen is an essential element needed in respiration, so a deficit can be lethal, but an excess is poisonous and can be just as debilitating as a deficit.

Hormones too have a profound influence on development. Of all the internal factors that influence behaviour, hormones are better studied and more fully understood than any other. There are two major reasons for this. First, hormone levels in the blood can be precisely measured and related to behavioural changes. Second, the endocrine organs that secrete hormones can be removed relatively easily. It is then possible to observe the effect of a particular hormone by studying behaviour (a) in the absence of the hormone (for those hormones whose absence is not lethal) and (b) after the injection of known amounts of it, in effect mimicking the function of the removed organ. The behavioural effects of testosterone, for example, are commonly studied by injecting it into animals that have been castrated (i.e. had their testes removed).

☐   Why do you suppose that this procedure is used? Why is it better to study the effects of testosterone on castrated, rather than on intact animals?

■   Because the amount of hormone present in the animal can be controlled. In an intact animal, the testes secrete testosterone in bursts and it would be necessary to take frequent blood samples to measure testosterone levels. If the testes are removed and testosterone is given by injection, how much testosterone is present in the animal's blood is known precisely, at least immediately after the injection.

This experimental technique has been used to investigate the influence of testosterone on the development of male sexual behaviour in the rat. The sexual behaviour of male rats can be measured in terms of which behaviour patterns are performed in the presence of a receptive female and how many times each of them is performed. (The three main components of male sexual behaviour are mounting, intromission (penis insertion) and ejaculation.) Female rats can sometimes be induced to perform mounting behaviour after injections of testosterone.

Typical results from experiments where the amount of testosterone was manipulated at birth and in adulthood in males are given in Table 5.1.

Table 5.1   Effect of the removal of testes at birth and the injection of testosterone on the performance of male sexual behaviour as an adult male.

	Manipulation at birth	Adulthood	Male sexual behaviour
1	Testes removed	testosterone no testosterone	none none
2	Testes removed + testosterone injection	testosterone no testosterone	present none

In experiment 1 all the rats had their testes removed at birth. As adults, half these rats received an injection of testosterone and half did not. In experiment 2 all the rats had their testes removed at birth and also received an injection of testosterone. As adults half these rats received a further injection of testosterone and half did not.

☐   What conclusions can be drawn about the influence of testosterone on the sexual behaviour of the male rat?

■ Experiment 1 shows that if testosterone is absent at or around the time of birth then, irrespective of whether testosterone is present or absent in adulthood, adult male sexual behaviour is not shown. Experiment 2 shows that if testosterone is present at or around the time of birth, then male sexual behaviour will only be displayed if testosterone is also present in adulthood.

These two experiments demonstrate two distinct effects of testosterone on behaviour. On the one hand there is a *priming* or *organizing* effect at or around the time of birth, and on the other there is an *activating* effect in adulthood. The organizing effect can be thought of as controlling differentiation of the nervous system, making it ready to respond to appropriate stimuli in adulthood. Interestingly, the organizing effect is graded—the extent of differentiation depends on the amount of testosterone present at this time. And the extent of differentiation affects how much male sexual behaviour the adult male displays. A large number of experiments have shown that there is a graded organizing effect of testosterone on female rats. In one set of experiments, the source of testosterone was their male litter-mates. In the normal litter there are four or five fetuses in each of the two parts (called horns) of the rat uterus. If one uterine horn is occupied exclusively by females, and those females are injected with testosterone when adult, then they will display only a very low level of male sexual behaviour (i.e. mounting), if any. In contrast, where females are next to males in the uterus, and those females are injected with testosterone when adult, they perform relatively high levels of mounting. The mere proximity of a testosterone-secreting male fetus in the uterus is sufficient to have an organizing effect on the behaviour of females.

5.3.2 External stimuli

The developing organism is subject to a continuous barrage of external stimuli from a very young age, for instance temperature and levels of oxygen and carbon dioxide are important from conception. Stimuli from touch and movement are also present before hatching or birth. It is difficult to assess the relevance of such stimuli to the normal development of the organism, but their presence should not be ignored.

Young bobwhite quail (*Colinus virginianus*) emit clicking noises while still in the egg. In this species the young are precocial and leave the nest to follow the mother soon after hatching. If there were only one chick this would present no problems, but the bobwhite quail lays a large clutch. How does the hen quail ensure that when she leaves the nest all the eggs have hatched? This is where the clicking noises come in. About two days before a chick is due to hatch it starts to emit clicks at the rate of between 80 and 150 per minute. In the nest the eggs are in contact and hatch within a few minutes of each other. If the eggs are separated so that they cannot communicate by clicks, the hatching is spread over a period of two days. The clicking therefore serves to synchronize hatching, either by speeding up development of the last laid eggs, or by slowing hatching behaviour of the first laid eggs, or some combination of the two.

After birth or hatching the range of possible external stimuli increases with olfactory and visual stimuli being added to the list. However for some species, these may not be significant straight away as the relevant sensory systems may be too immature to respond (e.g. young kittens are born with their eyes shut). This very immaturity can make the system extremely susceptible when it is eventually exposed to external stimuli. (This will be discussed in more detail in Book 4.)

The study cited in Section 3.5.2, in which maze-bright and maze-dull rats were reared in different environments is an example of external stimuli having a general effect on the development of behaviour, in this case performance in a maze.

The stimuli provided by conspecifics can be extremely varied and subtle and the effects may be very profound. For example, whether locusts develop into the solitary or the swarming type depends upon the amount of social contact the young locust makes. In the 1950s, Ronald Melzak raised puppies in the absence of conspecifics and found profound differences from normal puppies not only in social and sexual behaviour, as might have been expected, but also in their response to noxious stimuli. Whereas group-reared puppies responded to noxious stimuli (e.g. a lighted match or a pin) by moving smartly away, the solitary reared puppies did not attempt to avoid the heat of the flame or the pin however many times it was presented, and did not appear to experience pain sensations.

The effects of early social deprivation in primates were studied during the 1960s by William Mason and by Harry Harlow. By rearing infants in complete isolation, the initial studies imposed extreme conditions of social deprivation on newborn rhesus monkeys (*Macaca mulatta*). Monkeys reared in this way exhibited minimal exploration, considerable timidity, rocking behaviour and sometimes self-mutilation. However, not only did these infants not have conspecifics with which to interact, they also had an impoverished environment. Rearing monkeys with inanimate 'mother substitutes' (a head and body of towelling covering a wire frame) was sufficient to increase the amount of exploration and, if the inanimate objects were made to move, e.g. by attaching wheels and a motor, then there was even more exploratory behaviour and much less rocking.

These are clearly extreme conditions and it is perhaps not surprising that abnormal development ensued. However what is surprising is that the effects of complete isolation can be alleviated to a great extent if, during the period of isolation the young monkey is allowed just 15 minutes' interaction with its peers every day. This is a vastly reduced level of social interaction yet it is sufficient to decrease the effects of extreme deprivation considerably. On the other hand, noticeable effects on behaviour can result from very short periods of isolation in infancy. Robert Hinde, working in Cambridge, showed that five- to eight-month-old rhesus monkeys that were separated from their mothers for six days subsequently approached strange objects less and played less often than infants who had not been separated. These differences were present many months later.

External stimuli influence development in complex ways, and there is no clear relationship between the magnitude of differences in external stimuli and the magnitude of the behavioural differences that result.

5.4 Innate behaviour

As discussed in Section 1.2.3, the words 'instinctive' and, to a lesser extent, 'innate' have become devalued because they have been used in several quite different ways, in both scientific and everyday contexts. In this course, the term 'innate' is used for behaviour patterns that are performed correctly the first time they are performed during an animal's life, without prior experience or practice apparently being necessary. This definition is often taken to imply that such behaviour patterns are genetically determined but, as discussed in Section 1.2.3, this is an unjustified and erroneous assumption.

5.4.1 Species-specific behaviour

Some behaviour patterns are only shown by a particular species and are so predictable as to be definitive of that species. Here are three examples. The first illustrates a species-specific sequence, the second illustrates species-specific roles and the third, species-specific products.

Mating in the three-spined stickleback (*Gasterosteus aculeatus*) begins with the male performing a zigzag dance when the female appears (see Chapter 2, Figure 2.12). The dance attracts the female's attention, and, as she lingers, the male leads the female towards the nest he has constructed on the river bed. The male partly enters his nest, then backs out allowing the female to enter. He remains at the entrance and vibrates his body, during which time the female spawns. After spawning, the female leaves the nest, and the male enters and fertilizes the eggs. This behavioural sequence has some similarities to the mating sequence of the ten-spined stickleback (*Pungitius pungitius*), but also many differences; their behaviour is as characteristic of the species as is the number of spines.

The relative contributions of males and females to the rearing of the young varies consistently from species to species as does the amount of care given to the young. Female sheep pay considerable attention to their newborn lamb but sows all but ignore their piglets. Female cats lick and tend their offspring virtually continuously yet female rabbits, with equally helpless young, tend them for but an hour a day. In many species of bird (e.g. geese) the male and the female both tend the young whereas in others (e.g. moorhen, *Gallinula chloropus*) the male does most of the work, and in still others (e.g. domestic hens, *Gallus gallus domesticus*) it is left to the female.

Objects constructed by animals are often characteristic of the species, whether it be termite mounds, spider webs or birds' nests. The size, shape, composition and position of the nest identifies the builder. Wasps of the genus *Paralastor* dig an underground chamber and then construct an elaborate chimney over the entrance to it (Figure 5.3).

The songs and calls of many species of animals are instantly identifiable and can also be considered as species-specific.

Such a catalogue of species-specific behaviour patterns, all of which appear to be correctly performed on first occurrence and without practice, supports the notion that these behaviour patterns result from the growth and development of the nervous system in the absence of the relevant external stimuli and in the absence of feedback from muscles; that in some way the behaviour is 'programmed' into

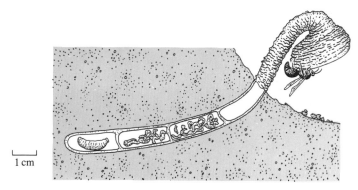

1 cm

Figure 5.3   A *Paralastor* chimney.

the nervous system. However, there are three reasons why such a notion is unacceptable. First, it is extremely difficult to identify all external stimuli that might affect the development of the nervous system and hence a particular pattern of behaviour. Unless all external stimuli can be identified, and ruled out as developmentally irrelevant, it remains possible that some external stimuli influence the development of the particular behaviour pattern. For instance, ducks may learn something of their mother's call while still inside the egg, and thus follow the noise of their mother preferentially.

Second, it is equally difficult to rule out the possibility that practice has occurred. In the study of pigeons reared in tubes cited in Section 5.2.1, the restrained birds may well have practised in an isometric manner—tensing and relaxing the appropriate muscles without any actual movement taking place.

Third, innate patterns of behaviour are not immutable, and individuals can, and do, change the way they perform those behaviour patterns over time. Laughing gull chicks (*Larus atricilla*) will peck appropriately at an object similar to the beak of their parents. However, the accuracy of the peck and its orientation improve with practice. A chick also learns only to respond to the beaks of its parents. Similarly, a young squirrel can open a nut but an older squirrel is more proficient at doing so.

One final point on species-specific behaviour patterns needs making. The fact that a behaviour pattern is species-specific does not mean that each individual performs that behaviour pattern in exactly the same way. A robin's song may be identifiable as a robin's song, but it is also different from any other robin's song, just as human faces are identifiable as human faces, and yet each is unique.

5.4.2   Species-specific behaviour and the stable environment

The species-specific behaviour patterns discussed in Section 5.4.1 appear generation after generation in a typical form. What could be more natural than to assume that such behaviour patterns were largely determined by the genotype of the organism? However, such an assumption ignores the large part played by parental behaviour in many species. Laying eggs in animals as varied as insects, amphibians and birds is an exact process. The environment into which the egg is put is not random, but carefully chosen or made by the animal to meet a number of different criteria. The result is that eagle eggs are laid in eagle nests, that holly leaf miner (*Phytomyza illicis*) eggs are laid in holly leaves, that bees' eggs are laid in appropriate cells in appropriate hives. (Mammals retain the zygote inside the body with the result that the environment in which it develops is controlled.)

Thus, not only do organisms develop in an appropriate environment, but they also develop in an environment very similar to that of their conspecifics, of preceding generations and of succeeding generations. The interesting corollary of this is that many species-specific behaviour patterns could be as much the result of a stable early environment as of a stable genotype.

The wasp *Nemeritis canescens* is a parasite; it lays its eggs in a living animal (called the host) so that the larvae can feed on its tissues when they emerge. The host of this particular wasp is the Mediterranean flour moth (*Ephestia kuhniella*). As with many invertebrate parasites, the wasp is very loyal to the one specific host moth, which it detects by odour. There are two possible causes of this loyalty: (i) the wasp has receptors that respond exclusively to the specific odour of the moth; (ii) the wasp learns the characteristic odour of the moth as a growing larva. William Thorpe (1938), working in Cambridge, was able experimentally to rear the wasp larvae on the larvae of another species, the small wax moth (*Meliphora grisella*). He then tested the adult females in an olfactometer (Figure 5.4) to discover which odour they would respond to. The results are presented in Table 5.2.

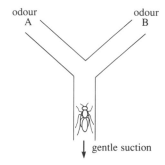

Figure 5.4 An olfactometer used by Thorpe.

Table 5.2 Percentage of female *Nemeritis* wasps approaching the odour of *Ephestia* or *Meliphora* larvae.

	Nemeritis raised on *Ephestia* larvae		*Nemeritis* raised on *Meliphora* larvae	
	Approaches *Ephestia*	Approaches *Meliphora*	Approaches *Ephestia*	Approaches *Meliphora*
Number approaching	154	27	389	202
Percentage	85	15	66	34

☐ Which of the two possible causes of loyalty do the data support?

■ The data provide evidence for both. Those wasps raised on *Ephestia* had a clear preference for *Ephestia* larvae as adults, though a small percentage did actually choose *Meliphora*. This pattern was retained in wasps reared on *Meliphora*, but a much higher percentage now preferred *Meliphora*. Clearly the choice of host is dependent on a predisposition to be responsive both to *Ephestia* and also to the host in which the larva develops.

Other experiments have been carried out to see whether a mother's diet could influence the food preferences of her offspring. Some rat mothers, for example, were fed a nutritionally adequate, but boring, commercial chow, while others were fed a highly palatable diet based on the protein casein. The pups were then tested for their preferred diet by presenting both diets simultaneously and recording how much of each diet they ate. The amounts of food consumed were small because, when tested, the pups were obtaining most of their nutrition from milk, but the pups whose mothers consumed the chow ate more chow than the other pups. Similar findings have been reported for a preference for garlic in rats in experiments in which a greater amount of garlic-flavoured water was consumed by rats whose mothers had consumed garlic-flavoured water while nursing.

Ducklings have been found to be preferentially responsive to the calls of their own species. Peking ducklings (a domestic variety of the mallard, *Anas platyrhynchos*) will follow a sound source emitting a conspecific call in preference to calls of other species. Again the question can be put as to the cause of this fidelity: it might be the result of the ears being especially sensitive to that kind of sound, or it might be the result of preferences acquired in the egg. Gilbert Gottlieb (1971) has demonstrated that ducklings do respond to conspecific calls while still in the egg by bill clapping and also that auditory isolation delays discrimination of the species typical call. He also found that ducklings needed to hear either their own noises or the noises of other ducklings to recognize the calls of their own species. Thus auditory stimulation influences the ducklings before hatching and affects their subsequent behaviour.

A more extreme example of how the environment in which development was successful is perpetuated is given by indigo birds (*Vidua chalybeata*). These weaver finches are brood parasites, laying their eggs in the nests of related species rather than weaving their own nests. The open mouths of the nestlings of the host species present a particular array of stimuli to the parent birds, often with dots or lines inside the mouth. Only this particular stimulus array will cause the parent bird to go through the elaborate process of regurgitating food directly into the throat of the nestling. The brood parasite, therefore, has to have the correct host species stimulus array, otherwise its nestling will not be fed. Furthermore, because indigo birds are behaviourally polymorphic, with different indigo birds laying their eggs in the nests of different host species, it is important that only indigo birds raised by the same host species should mate together, otherwise the feeding stimuli of the progeny would be wrong. The first condition is met by the female learning the characteristics of the host nest as a nestling, and subsequently laying her eggs only in similar nests. The second condition is more difficult to meet. How can birds of the same species (i.e. indigo birds) distinguish those conspecifics raised by one host species from those raised by another host species? The answer is in song. Young indigo birds learn their songs from their foster parents. Later the males sing songs, and females choose to mate with males singing a familiar song. Thus mating is always between birds that have had the same species as foster parents.

What these three examples illustrate is that species-specific behaviour is affected by external stimuli. There is no absolute distinction between a behaviour pattern that is species-specific and other behaviour patterns, it is just that some behaviour patterns are influenced more by the environment than others. Under normal circumstances developing organisms experience a very similar environment to that of their parents when their parents were developing, and also to that of conspecifics. Thus although the development of behaviour may be labile, this is usually masked by a relatively constant array of external stimuli.

5.5 Imprinting

No introduction to the development of behaviour would be complete without reference to **imprinting** because imprinting is a special form of learning with unique characteristics. Essentially, imprinting is the process whereby some young precocial animals form an attachment to, and follow, a large moving object in their environment. Imprinting (sometimes referred to as filial imprinting) occurs in a very short space of time, very early in independent life (i.e. soon after hatching, or birth) and under normal circumstances the imprinted object would be the mother.

Filial imprinting is a phenomenon associated with precocial young. Precocial young normally become attached to their mother within the first day or two after hatching or birth, and follow her but avoid other moving objects. Imprinting ensures that the young stay close to a source of protection and also, in some cases (e.g. mammals such as sheep and antelope), a source of food.

Imprinting shot to fame in 1930 when the German ethologist Konrad Lorenz announced that he was foster parent to a family of goslings which had imprinted on him. Since then the range of stimuli to which young animals would imprint has been found to be vast. Contrary to expectation, the imprinting stimulus does not have to be animate or even vaguely animal like; the imprinting object can be a ball, a cube, or even a flashing light, though it typically must be within a particular size and colour range, and be moving, to be most effective. There are examples of zebra foals becoming imprinted on cars.

Imprinting, then, is another example where the environment in which the phenomenon normally happens is sufficiently stable, generation after generation, to ensure that the young normally imprint on the mother—she is the first large moving object the young normally see.

There are several questions that can be asked about imprinting. For instance, you could ask questions about the timing of imprinting: for example, at what stage in development does it occur, how long does it take and how long does it last? You could also ask about the quality of the imprinting and the ability of the young animal to discriminate between the imprinting stimulus and some other stimulus. Alternatively, you might be interested in the processes that underlie imprinting (imprinting is often studied as a special case of learning and is considered in this context in Book 4). The experimental procedure used to investigate all these questions is the same, so the basic procedure is presented before these questions are considered.

5.5.1 Experimental imprinting

Most studies of imprinting have concentrated on the visual stimuli presented to the young animal. Although visual stimuli alone are quite sufficient to induce imprinting, it should be remembered that the mother presents a fairly impressive constellation of auditory, olfactory, tactile and visual stimuli.

Precocial birds (chicks or ducks) are the usual subjects in studies of imprinting. The newly hatched bird is removed from the presence of conspecifics and kept where it has the opportunity to approach, nestle against or follow another object.

This object may be anything from an abstract shape, to a realistic model (including noises and warmth) of its mother. The amount of exposure the young bird has to the imprinting stimulus can be varied from minutes to days, can be continuous or discontinuous and can begin any time after hatching. At some later time the young bird can be tested for its preference between the imprinting stimulus and some other stimulus, or for its ability to discriminate the imprinting stimulus from some other similar stimulus. A common design of apparatus is shown in Figure 5.5.

Figure 5.5 Apparatus used in imprinting experiments. The imprinting stimulus is moved around the arena mechanically. During the imprinting period there would only be one stimulus in the arena, attached to the end of one of the arms. During later testing the imprinted stimulus would be on one arm and a control stimulus would be on the other arm.

5.5.2 The timing of imprinting

The image of a young bird, struggling out of its shell, opening its eyes and instantly imprinting on its mother is powerful, but wrong. The question of when imprinting occurs was studied by Eckhard Hess in Chicago in 1959. He used the apparatus in Figure 5.5 and allowed mallard ducklings of various ages something less than an hour in the presence of the imprinting stimulus, an adult male of the same species. An hour later the ducklings were tested and scored as to whether they approached the imprinting stimulus, a model adult female of the same species (the male and female have strikingly different plumage), or made no choice (Figure 5.6).

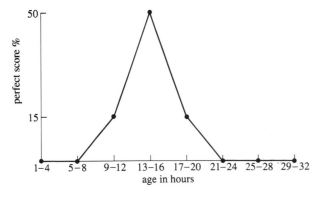

Figure 5.6 Percentage of ducklings of different ages which approached the imprinting stimulus one hour after first being exposed to it.

☐ What do the data in Figure 5.6 reveal about when imprinting occurs?

■ Essentially these results show three things: (i) imprinting does not occur immediately post-hatching; (ii) there is a period of about 12 hours during which imprinting can occur; (iii) after a certain age imprinting does not occur.

In the first few hours after hatching, chicks are not very coordinated, teetering and tottering somewhat unpredictably. So the first following response must await the development of balance and coordination. Also, there could well be a need for other sensory systems to mature. Together these two factors would account for the absence of any response in the first few hours after hatching.

The period of 12 hours during which the ducklings were particularly sensitive to imprinting stimuli is called a *sensitive period*, a very important concept in developmental biology.

Sensitive periods also illustrate the importance of distinguishing between age from birth or hatching, and age from conception. For instance, one criticism of Hess's study was that he used post-hatching age rather than age from conception (i.e. *developmental* age). He thus ignored any differences in development that might arise from differences in the duration of incubation. Other experiments have eliminated the problem of variation in hatching time by testing Peking ducklings of similar post-hatching age (i.e. 11–20 hours). These show that imprinting is more successful between 650 and 669 hours of developmental age than at 640–649 or 670–679 hours of developmental age. In other words, even for birds of similar post-hatching age there was a sensitive period that related to the length of time the birds had been incubated before hatching. Thus differences in responsiveness arise as a result of differences in developmental age, as well as from differences in age after hatching.

The end of the sensitive period is even more difficult to determine. In the normal course of events, once imprinting has occurred the young animal is no longer sensitive to imprinting stimuli. So imprinting itself closes the time-window. To prevent this from happening and to test the sensitivity of older animals to imprinting stimuli it is clearly necessary to keep animals and stimuli apart. However, the problem with this is that in the absence of obvious imprinting stimuli the older animals imprint on any available stimuli, even the static cage in which they are reared.

☐ How does this information tie in with the data in Figure 5.6 where Hess quite clearly shows an end to the sensitive period?

■ The ducklings at 24 hours or more of age may not have followed the imprinting stimulus because they were already imprinted on something else. Thus what appears to be the end of the sensitive period really marks the time it takes for the animal to imprint on whatever stimulus is available.

5.5.3 The role of the young

Imprinting is not a one-way process in which external stimuli impinge on the passive young bird. It is a two-way process because the young bird actively seeks out external stimuli. A young duckling put into a featureless box will move around noisily. If an object is put into the box the duckling will run to it and stop calling. Furthermore, it is possible to train ducklings to press a pedal if the pedal switches on a flashing light—an object on which they will readily imprint. The activity of the young bird during imprinting is important because the young bird must learn to recognize its mother from a variety of different angles. In the absence of such activity the young bird might, for instance, imprint on the side view of its mother. When the mother moved away, the young bird would only see her rear view, not her side and thus not the imprinting stimulus. Without the imprinting stimulus the young bird would not follow her.

5.5.4   The function of imprinting

Despite the intensive study of the causes and parameters of imprinting, no one really addressed the issue of the *function* of imprinting until relatively recently.

☐   A possible function of imprinting was mentioned in the introduction to this section. What was it?

■   To keep the young close to a protector and possibly a source of food.

Timothy Johnston and Gilbert Gottlieb (1981) pointed out that, if imprinting is to have such functions, the young birds need to be able to distinguish their mother (as the imprinted stimulus) from any other bird in the flock, or on the pond. Somewhat disturbingly Johnston and Gottlieb were unable conclusively to demonstrate that mallard ducklings could perform this feat. Although the ducklings imprinted on mallard mothers could distinguish them from pintail ducks (*Anas acuta*), they could not distinguish them from red-headed ducks (*Aythya americana*). And as differences between species are greater than those within species this should have been a relatively simple task. The question remains open, but it is salutary to reflect on how easily function can be forgotten.

5.5.5   Mate preferences

Where young have experience of their parents, it is quite possible for that experience to influence the eventual choice of mate by the young. The young learn the essential characteristics of the species from their parents and then, when the time comes to mate, they seek such characteristics in a partner. The process of learning the species characteristics is called **sexual imprinting**.

Like filial imprinting, sexual imprinting can lead to animals forming attachments for 'inappropriate' objects. Birds that are reared by a foster-species commonly form a sexual preference for their foster-species rather than their own, and Lorenz's geese attempted to mate with him. In many sexually dimorphic species, such as the mallard, in which the male is brightly coloured, only the female cares for the young.

☐   What consequence will this have on sexual imprinting for male and female ducklings?
■   If male ducklings sexually imprint on their mothers they will subsequently form a sexual attachment for females of their own species, and so will mate with members of their own species. If female ducklings sexually imprint, however, they would tend to form homosexual attachments with other females.

In sexually dimorphic species, in which only one parent cares for the young, sexual imprinting is thus typically confined to one sex. The other sex requires some other mechanism for identifying sexual partners correctly.

Recent research indicates that sexual imprinting can, in fact, be rather more specific than was previously thought. Japanese quail (*Coturnix coturnix japonica*) develop a sexual preference for individuals with plumage characteristics that are similar to, but not identical to those of their immediate family. As a result, when they come to form pairs they tend to avoid mating with very close relatives, such as their siblings, but are more likely to mate with more distant relatives, such as their cousins, than with quite unrelated individuals.

The major feature that sexual and filial imprinting share is the sensitive period. But whereas filial imprinting occurs in the first few days of life, sexual imprinting does not begin until some weeks after hatching and lasts for many weeks. Furthermore, sexual imprinting is not evident until the animal reaches sexual maturity and starts behaving differently towards males and females. Up to then sexual imprinting is latent. Sexual imprinting also shows *generalization* in that the sexual responses of the imprinted animal are directed towards members of the species on which the animal is imprinted, rather than towards a particular individual. This contrasts with filial imprinting where following responses may be directed towards a particular individual.

Sexual imprinting was clearly revealed in the cross-fostering experiments carried out by the German ethologist Klaus Immelmann in the early seventies. In these experiments, birds of one species (a) are reared by another species (b), while birds of species (b) are reared by birds of species (a). When sexually mature, birds are tested for their preference for a mate between birds of the same species as themselves and birds of the same species as their foster parents. A preference for foster parent look-a-likes confirms that sexual imprinting has occurred. This experiment was performed using Bengalese finches (*Lonchura striata*) and zebra finches (*Taeniopygia guttata*). A male zebra finch raised by Bengalese finches will, when sexually mature, direct its attention to a female Bengalese finch, even in the presence of a female zebra finch. This is despite the fact that the zebra finch female directs her attentions towards him, while the female Bengalese finch ignores him. If a male zebra finch is raised by a mixed pair of Bengalese and zebra finches, he will imprint on the zebra finch, irrespective of whether the zebra finch adult is a male or a female. Clearly the young bird is more receptive to some stimuli than to others.

☐ There is another possible explanation of this latter result that has to do with the behaviour of the parent birds. What is it?

■ The parent birds may recognize their charge as belonging to the same or a different species and alter their behaviour accordingly. The young bird may then imprint on the more attentive parent.

Species differences in sexual imprinting raise a number of questions about the evolution of behaviour. In the mallard, only the adult female shows parental behaviour, and only the males show sexual imprinting. In the Chilean teal (*Anas flavirostris*), both sexes show parental behaviour and both sexes can be sexually imprinted. How did these differences evolve and what is their function? Answers to these questions await further study.

5.5.6 Imprinting on non-visual stimuli

Filial imprinting is essentially an acquired preference for the visual stimuli presented by the mother. Now there are a number of other stimuli for which a developing organism could form a preference, sometimes a lasting preference; this would be a phenomenon comparable with imprinting. For instance, the call of adult guillemots (*Uria aalge*), a sea-bird that nests in dense colonies, varies from individual to individual. Young guillemots will move towards the call of one of their parents in preference to the call of another guillemot. This discrimination is displayed by newly hatched chicks. If chicks are reared in incubators and played calls before hatching, when the chicks hatch, they will preferentially move towards calls that have been played to them while in the egg. This has been called auditory imprinting.

Smell is another stimulus for which animals develop early preferences. In the precocial spiny mouse (*Acomys cahirinus*), if the young are exposed to the smells of spices such as cinnamon or cumin in their bedding then they demonstrate a preference for that odour a short time later. Similar olfactory imprinting has been found in the preferences of young human babies for milk. When given the choice of turning their head towards the smell of their mother's milk or towards the smell of another woman's milk, the babies turned towards their mother's milk.

Taste, or food preferences in young animals were mentioned earlier; young rats preferred the food their mothers had eaten. If six day old herring gulls (*Larus argentatus*) are offered a choice of chopped worms, pink cat food or green cat food, the gulls prefer the food they have been fed on over the previous five days.

☐ What do these studies demonstrate?

■ These studies show that some preferences in some young animals are acquired in somewhat unexpected ways. The point is that in considering the development of behaviour these 'unexpected ways' are, by definition, easily overlooked, yet they may be crucial to an understanding of an animal's behaviour.

By the way, young animals need not *necessarily* acquire a particular food preference. For example the garter snake (*Thamnophis sirtalis*) gives birth to live young. Could the type of food the mother had been eating affect the food preferences of her newborn young? Newborn garter snakes show no difference in their responses to cotton swabs containing worm extract or fish extract according to which of the two diets their mother had been eating.

Perhaps the last word in the long lasting influence of preferences acquired in early life should go to the salmon (*Salmo salar*). Their extraordinary life begins high up near the source of a river. Over many years they migrate down the river and out to sea, sometimes swimming vast distances. After four to seven years at sea these fish manage to return almost exactly to the place of their birth, a remarkable feat, achieved by remembering and following the smell of their natal river.

5.6 Bird song

Bird song is a well studied example of behavioural development. It exemplifies many aspects of behavioural development with one conspicuous advantage, that it can be recorded and measured very accurately.

Singing in birds is a very serious business. It is mostly done by males and serves two main functions (though the balance between these functions may vary from species to species): first, to attract a mate and, second, to deter rival males. The song will only work, though, if the females that are attracted and the males that are deterred are of the same species as the singing bird, a feat which is accomplished by the use of species-specific song. The question that arises is how does an individual bird come to sing the song appropriate to its species?

The initial studies of William Thorpe with chaffinches (*Fringilla coelebs*) in the 1950s suggested that young birds need to be exposed to the song of adult males at appropriate times during development if the chaffinches are to sing species-specific song as adults. These studies continue to be refined by Peter Marler in the USA using white crowned sparrows (*Zonotrichia leucophrys*), song sparrows (*Melospiza melodia*) and swamp sparrows (*Melospiza georgiana*). The scheme of song learning proposed by Marler is presented in Figure 5.7.

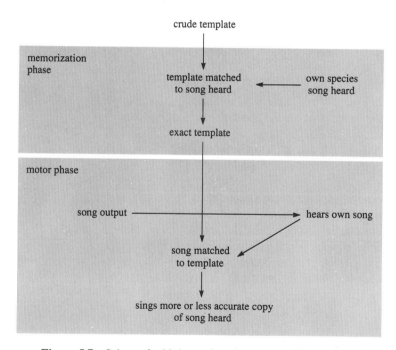

Figure 5.7   Scheme for bird-song learning proposed by Marler.

The bird hatches with what Marler calls a 'crude template'. (A template is something that can be used for comparison, like a paper dress pattern, though in this case it is an unspecified pattern in the nervous system.) There then follow two phases: a memorization phase, during which the young bird is silent but listening to the sound of its own species song; and a motor phase during which the young bird practices its own song.

The evidence for the template is that birds denied all auditory stimulation, including any noises they make themselves, by being deafened, just make noises (a screech). The nerves that stimulate the syrinx (the organ in the bird's throat that creates the noise) are clearly present and active, but the result is unrefined. However, if birds are denied all auditory stimulation except their own, they produce a characteristic rudimentary song. The important point here is that this rudimentary song is different for different species and is the product of an unrefined template.The rudimentary song matches the crude template and the bird continues to sing the rudimentary song.

The evidence for the memorization phase is that if the young sparrow hears an appropriate song during its first 100 days of life, it then goes on to sing full adult song the following year. If it does not hear the song during this period, either by being kept in isolation, or by being deafened, it sings a very poor song. In Marler's scheme, hearing the song modifies the template so that the template exactly matches the full adult song. When the bird comes to sing as an adult, it matches its output to the refined template and, after a little practice, produces full song. The memorization phase is over by about 100 days of age, because if the young bird first hears the adult song after 100 days of age it will not modify its simple song, i.e. it has a sensitive period.

The motor phase begins in the spring following hatching, when the bird first starts singing. Its early attempts at singing are crude, with many additional notes, but as it practices and hears its own efforts, so it shapes its song output to match the template. The important point here is that the bird must hear its own song. If it is deafened at 100 days of age, so that it cannot hear its own song, it will sing only a rudimentary song, despite having a template for full song.

Although it is possible to hear the differences between full song and the song of isolated or deafened birds, the ephemeral and brief nature of bird song make detailed analysis difficult. Fortunately it is possible to use a machine, a sound spectrograph, to convert sounds into graphical form, called sonagrams. A sonagram can, like that shown in Figure 5.8, depict time along the horizontal axis and sound frequency (pitch) along the vertical axis. Thus, in a sonagram of a bird song, a black line runs from left to right for as long as the particular call lasts. The higher the pitch of the call, the higher up on the graph is the mark made. In this way it is possible to measure how long each note in a song lasts, to within a few milliseconds if needed, and to examine variations in pitch.

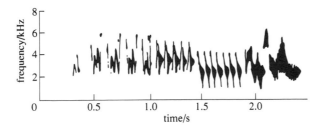

Figure 5.8 Sonagram of a chaffinch song.

With this apparatus it is possible to see (Figure 5.9, *overleaf*) the full song, rudimentary song and screech described above.

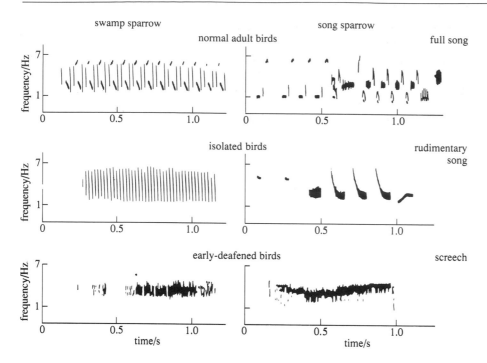

Figure 5.9 Typical songs of a male swamp sparrow and a male song sparrow, shown first in natural form then as produced by males reared in isolation from the song of their species in infancy, and as developed by males deafened in infancy. The songs of isolated birds are much simpler, but still to some degree species-specific, while this is not true of the screech produced by early-deafened birds which have heard neither other individuals nor their own voice.

☐ Compare the song of the isolated song sparrow with the song of the isolated swamp sparrow. Is there any evidence for a template?

■ Yes. If there were no template, there would be no reason for the songs of isolated birds to be different between the two species. Isolated birds have never heard their own species-specific full song, and yet they produce species-specific rudimentary song.

The development of bird song, like imprinting, has a sensitive period. But in contrast to the generalized stimuli which will elicit imprinting, young birds are sensitive to very specific aspects of bird song. Bird song is composed of discrete syllables which are joined together to make the song; species differ in the syllables they use and in the sequence in which those syllables are joined together. Swamp sparrows will only learn syllables of their own species, but will do so even if these are edited in such a way as to follow the pattern of a song sparrow song. However song sparrows are prepared to accept swamp sparrow syllables provided they are organized in song sparrow fashion. They will not learn normal swamp sparrow song. These birds, and others (e.g. the chaffinch) are extremely selective over what stimuli will influence the template.

This developmental scheme fits many species of songbird and, in the proper scientific tradition, the scheme allows comparisons between species to be made and exceptions noted. For instance the marsh wren (*Cistothorus palustris*) can learn a number of different songs (Figure 5.10), but the marsh warbler (*Acrocephalus palustris*) incorporates bits and pieces of song from numerous different species in its own song and the canary (*Serinus canaria*) continues to change its songs in adulthood. In the zebra finch young birds will not learn from loudspeakers or tape recordings and indeed may require visual as well as vocal interaction with the 'tutor' to learn song from them. These exceptions pose interesting questions about how and why species differ. Some of these questions are taken up later in the course.

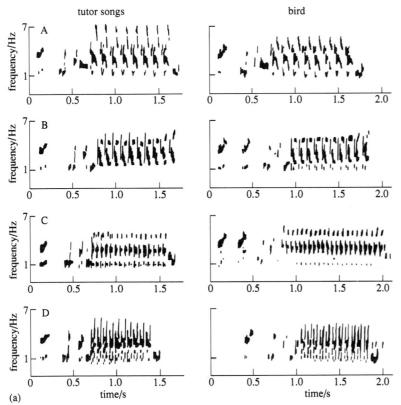

Figure 5.10   Sonagrams of (a) tape recordings of four songs played to a marsh wren as a juvenile and (b) the songs learnt by the bird as a result.

Summary of Chapter 5

A range of factors that influence the way in which an organism changes from a zygote to a behaving adult have been considered. Development is a continuous transaction between the genotype and the environment and the resulting behaviour cannot be exactly predicted. Similarities in the behaviour of animals of the same species arise because of similarities in their genotype and the environment in which they are raised. Differences in the behaviour of animals of the same species can result from differences in the precise stimulation received in the form of internal and external stimuli, the time the stimulation was received, and whether the stimulation was received before, during or after a sensitive period.

Objectives for Chapter 5

When you have completed this chapter, you should be able to:

5.1   Define and use, or recognize definitions and applications of each of the terms printed in **bold** in the text. (*Questions 5.1 and 5.2*)

5.2   Explain what is meant by a developmental path. (*Question 5.3*)

5.3   Give evidence that both internal and external stimuli affect the development of behaviour. (*Question 5.4*)

5.4   Give examples of and discuss factors that influence the development of species-specific behaviour. (*Question 5.4*)

5.5   Use examples to explain the significance of imprinting in behavioural development. (*Question 5.5*)

5.6   Discuss the relevant issues in the development of behaviour, using bird song as an example.

Questions for Chapter 5

Question 5.1 (*Objective 5.1*)
What is meant by the term 'sensitive period'?

Question 5.2 (*Objective 5.1*)
Most people behave and look like people. How can this be reconciled with the statement in Section 5.2 that 'development does not follow a blueprint or programme in some inevitable way'?

Question 5.3 (*Objective 5.2*)
In the study in which mice from two strains were cross-fostered at birth (Section 5.2.2), the mother was said to influence the aggression score of her foster pups. What kind of stimuli could the mother provide that might affect the young pups?

Question 5.4 (*Objectives 5.3 and 5.4*)
Define filial, sexual and olfactory imprinting.

Question 5.5 (*Objective 5.5*)
What is the evidence for the song template?

References

Cruze, W. W. (1935) Maturation and learning in chicks, *Journal of Comparative Psychology*, **19**, pp. 371–409.

Gottlieb, G. (1971) *Development of species identification in birds*, University of Chicago Press.

Hess, E. (1959) Two conditions limiting critical age for imprinting, *Journal of Comparative and Physiological Psychology*, **52**, 515–518.

Johnston, T. D. and Gottlieb, G. (1981) Development of visual species identification in ducklings: what is the role of imprinting?, *Animal Behaviour*, **29**, pp. 1082–1099.

Thorpe, W. H. (1938) Olfactory conditioning in a parasitic insect, *Proceedings of the Royal Society of London, Series B*, **126**, pp. 370–397.

Further reading

Halliday, T. R. and Slater, P. J. B. (eds) (1983) *Animal Behaviour 3: Genes, Development and Learning,* Blackwell Scientific Publications.

Johnston, T. D. (1988) Developmental explanation and the ontogeny of bird song: nature/nurture redux, *Behavioural and Brain Science*, **11,** pp. 617–663. (For those interested in pursuing this subject in more detail.)

CHAPTER 6
LEARNING

6.1 Introduction: what is learning?

We have used the term 'learning' earlier in this book without defining it. It may come as a surprise to you to discover that providing a satisfactory definition is rather difficult. You may realize some of these difficulties if you spend a few minutes listing things you have learnt in your life.

Your list probably contained a diversity of activities from learning to walk and talk, to tying a shoe lace, riding a bicycle or playing chess, as well as the kind of learning associated with reading this book and the learning of social skills, such as which topics of conversation to avoid when in the company of a particular neighbour. This is quite a heterogeneous group of behaviours but in every example the outcome is that some new knowledge or skill has been acquired.

Think now about animal **learning**. Do any animals have to learn skills such as walking? Chapter 5 described experiments which showed that pigeons do not learn to fly by practicing wing flapping. This reminds us that it is unwise to *assume* that a change in behaviour is a result of learning. However, in the same chapter you read that young squirrels can open nuts but older squirrels are more proficient. Experiments have shown that this improvement in handling technique is learned. You have also read about more complex learning in Chapter 2 where domestic chicks formed search images and it was suggested that they 'learnt to see' the food. In the same chapter, experiments with the digger wasp showed that the female used local landmarks to re-find her nest entrance after a short absence. Experiments like these suggest that the ability to learn is widespread.

All these examples suggest that human and animal learning abilities are similar; skills can be learned and knowledge assimilated.

Most animals can learn certain things but some animals show very special learning abilities; a few examples are imprinting (Chapter 5), song (Chapter 5) and food hoarding. Many animals store food when it is abundant and then return to these hoards later when food items are scarce: squirrels hoarding nuts might seem the obvious example to give here. Less well known are the 'worm larders' of the mole. Moles bite and paralyse worms before eating them but any worms caught surplus to immediate requirements are stored in a 'larder' within the mole's network of runs. The worm is paralysed and cannot escape but it is alive and so provides a fresh meal for the mole when required. Another food storing animal is the Canadian bird, Clark's nutcracker (*Nucifraga columbiana*). It has a strong sharp bill that enables it to prize pine seeds out of cones before they open and a throat 'pouch' that can carry many seeds at once. These seeds are stored in several locations on mountains that remain free of snow in winter. The birds raise young early in the year using the seed hoards to provision themselves and their offspring. Animals such as these seem to show considerable feats·of learning and memory, but can we be sure that learning has occurred?

☐ How else could the food hoards be relocated?

■ The squirrels and birds may simply have 'preferred areas' that they visit regularly and systematically search. The mole may relocate the worms by smell.

Even if learning has occurred it may be that the animal remembers the general area rather than the specific site.

It is quite difficult to carry out investigations in the field to establish what type of experience the animal needs in order to learn. For example, does the animal need to store the food itself, or, if it finds other caches or sees a conspecific hoarding food can it remember the locations of these food stores just as easily?

In Chapter 2 it was explained that the way in which animals respond to the environment can often be understood if we know how they perceive that environment; in addition to the effect that they have upon an animal's *current* behaviour, environmental events (stimuli) are, in general, able to influence its *future* behaviour. This is possible because, in some way which for the most part is still unclear, particular events that impinge upon the animal's senses and their associated consequences are coded or registered in the nervous system. For instance, an animal touching a hot object experiences the consequences and, in future, tends to avoid that object. You see the animal avoiding the hot object and say that learning has occurred. In this case learning has been recognized because behaviour has been modified as a result of experience.

In general, there is so much happening in an animal's environment that it is impossible to know what knowledge it is assimilating at any one moment. Learning is occurring continuously but it does not always change behaviour immediately.

☐ Can you think of an example of learning that might not be manifested by changed behaviour?

■ You perhaps thought of the learning process you are now undergoing. It is possible that imprinting is a new topic about which you have recently learned. However, unless you are asked to give an account of this phenomenon it may be that you will never speak or write about it. There will be no behavioural evidence of your learning.

Learning can only be revealed in *performance*, which in turn depends on **recall**, defined as the successful use of learned experience, or the expression of a response modified as the result of learning. When investigating learning it is obviously very important that appropriate tests of recall are made. For example, a badly worded exam question may elicit an unexpected response. The examiner has then failed to discover whether or not the students had learnt about a particular topic. It is equally difficult to devise animal experiments that provide unequivocal evidence of learning.

Often a considerable time elapses between learning and recall. It is believed that as a result of the learning process there is a change in the nervous system—a **memory**. Memory can be defined as a change in the nervous system consequent on learning, by which information is stored—it is the inferred intervening process that connects learning to recall. (The evidence for this change in the nervous system will be presented in Book 4 after you have read more about the nervous system in Book 2.) Thus **learning** could be defined as the process by which a memory is acquired. Memory, like learning, cannot be observed directly; we infer it from behaviour.

Recall depends on both:

1   The extent to which the material was originally stored in memory. If an animal is unable to recall something it could be because, on original exposure, it had insufficient effect, i.e. learning did not take place.

2   The extent to which the material is still available. If an animal is unable to recall something it could be that the effect of the original exposure has worn off (been forgotten) or has been actively discarded ('unlearning' has taken place), or because 'access' to the memory is blocked for some reason.

In considering these possibilities the timing of events may be important. For example, the information may be irrelevant at the time and therefore not learned. You could easily have learned the surname of those pleasant people you met on holiday but you only heard it once and they lived miles away from you. Two years later when you move to the same town you wish you could look for their name in the phone book. Timing may be important in another way. A piece of information might not have been used for years and may be judged to be of no further use, such as the phone number of your best school friend whose parents have long since left the neighbourhood. Have you lost the number from your memory, i.e. is it no longer available, or are you just unable to recall it?

The distinction between loss of memory and failure of recall will be considered further in Book 4. It is mentioned here to highlight the difficulty in interpreting experiments on learning. Inability to demonstrate learning may not prove that the animal failed to learn.

Figure 6.1 illustrates the learning process. Stimuli impinge upon the nervous system, and the animal's current behaviour is determined by the way in which the nervous system processes the incoming information. However, there may not be a behavioural output that the observer can identify as being the result of the learning process and so there is no simple, universally accepted definition that encompasses all examples of learning.

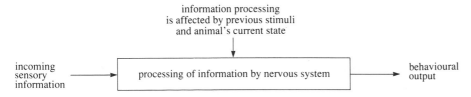

Figure 6.1   The relationship between current and past stimulation.

In general, animals learn about events in their external and internal (bodily) environments, and about relationships between these events. For instance a blue tit (*Parus caeruleus*) learns that pecking through the foil of a milk bottle yields food (a relationship between two external events) and a rat learns that drinking salt-water increases dehydration (a relationship between an external and an internal event). What is learned and how easily it is learned varies between individuals. It also varies from species to species, as will be shown in Section 6.4. Different species have different environments and different evolutionary histories. They are morphologically and physiologically different, and it is therefore unlikely that their learning abilities will be the same.

Learning occurs most readily when it has consequences for the animal's fitness. For example, a hungry blue tit finds a novel food and eats it. The food provides nutrients and therefore the discovery has a direct consequence for the animal's physiology. It may therefore learn not only the site of the food but an association between the taste of food and its beneficial consequence.

The same bird may repeatedly fly into a closed window before learning to avoid it, a task that to the human observer may appear simple to learn. Section 6.4 will consider constraints on learning in more detail.

☐   What is the biological advantage of being able to learn and remember?

■   These abilities provide flexibility. If behaviour is based not only upon present, but also upon past stimuli, an animal is better able to survive. For example, if an animal learns that a certain route leads to food, it saves considerable time when it is subsequently hungry. Suppose the animal's staple diet is scarce and it samples new foods. These may have either beneficial, neutral or harmful consequences. If the animal bases its choice— to ingest or avoid—upon past experience, it will be more likely to survive.

☐   What, if any, is the alternative to having the ability to learn?

■   An alternative strategy is having fixed, highly stereotyped behaviour adapted to specific environmental and bodily conditions. If circumstances change rapidly, this may leave the organism vulnerable. For example, what does it do when the staple diet is no longer available? Suppose it tries an alternative food and this has harmful consequences. It would have no means of avoiding this food in future if it is unable to learn.

6.2   The study of learning: different approaches

Different groups of scientists have traditionally approached the study of learning in different ways, looking at different aspects of the problem. In this chapter we will consider contributions from ethologists and psychologists, both groups of researchers being interested in behavioural measures of learning. These two groups have to be treated separately because, historically, there have been misunderstandings and disagreements between protagonists of the two approaches. Today, despite far better understanding of each other's position and more cooperative research projects, the two groups still pursue slightly different aims.

The primary concern of ethologists is to study the behaviour of animals in their natural environment. Consequently, they are primarily interested in the function that learning plays in an animal's life, and in species-specific differences in learning abilities. They have also been interested in the role of learning in development (see Chapter 5). They have tended not to look for general mechanisms underlying learning. This may be because learning abilities seem often to be so specific to the task and to the species as to defy any attempt at generalizations. If you consider what you now know about the learning involved in producing species-specific bird song (Chapter 5) you might wonder why these birds do not show evidence of equally sophisticated learning in other areas of their activities. Also, why are there species differences such that some birds are quite constrained in what they can learn whilst others are not? Again, if the bird is learning about song, an auditory stimulus, why should it learn better if the tutor is another bird rather than a tape of the song? Thus the complexity of the task as well as the specificity has left unanswered questions as to how this learning is achieved.

The primary concern of psychologists in studying learning is to elucidate the processes that underlie it. They have tended not to be concerned with species differences; indeed, their aim is to identify mechanisms of learning that are common to all animals. By looking at simple examples of learning and studying them under rigidly controlled conditions, psychologists have sought to provide general laws of learning. This approach has been relatively successful but recent work now questions the universality of some of their findings (see Section 6.4.2).

Other scientists who study learning and memory are neurophysiologists and biochemists. Their contribution is beyond the scope of this book but will be considered in later books. The first examples of learning to be described in this chapter, habituation and sensitization, have been much studied by neurophysiologists, so you will be using this information again in Book 4. In this chapter only the behavioural aspects will be considered.

6.3 Types of learning

Three main categories of learning can be distinguished:

1 **Non-associative learning**, in which an animal is exposed repeatedly to a single stimulus.

2 **Associative learning**, in which an animal is exposed to two or more stimuli that have a particular relationship to each other. The animal demonstrates associative learning if it shows evidence of having made the association between the two stimuli.

3 **Complex learning**.

The differences between these categories of learning should become clear in the following sections as they are discussed in detail. There remains considerable debate as to whether these distinctions reflect different processes underlying the behavioural changes that are observed. In other words, there is disagreement as to whether these processes can all be classified as 'learning', a unitary process, with general laws governing its acquisition.

6.3.1 Non-associative learning

One form of non-associative learning, called **habituation**, is a simple form of learning which seems to be ubiquitous. It is a reduction in the effectiveness with which a stimulus elicits a response, if the stimulus is repeatedly applied with no serious consequences to the organism. For example, advertisers keep changing advertisements because they assume we habituate to them. If you lean over a rock pool and brush a sea anemone's tentacles lightly with a delicate object such as a paintbrush, it will withdraw its tentacles. However, if you repeat the process each time the tentacles are again extended there will be a progressive diminution of the response until, eventually, the tentacles will not be withdrawn at all. (You will need a lot of patience, sea anemones are relatively slow to extend their tentacles once they have withdrawn them.) Those of you who cannot find a convenient rock pool could try something similar with a snail. Lightly tap on the ground beside a moving snail and see what happens. As you would expect, it will withdraw into its shell. When it has emerged and started to move again, tap the ground again. If you repeat this procedure you will notice two things. First, that the snail spends progressively less time in its shell after your tap and second, that it does not completely withdraw into the shell anyway. In fact when you tap the ground the fourth or fifth time you may not get any response from the snail.

☐ What is the benefit to the animal of this kind of behaviour?

■ It saves energy and time. For example, it might be a harmless piece of seaweed that is brushing the sea anemone's tentacles and the tap on the ground beside the snail might be ripe fruit falling from a tree. If animals continue to respond to non-noxious stimuli they use unnecessary energy; whilst responding they also waste time which could be spent on more profitable activities, like eating.

Extensive experimental investigations have been carried out on habituation. Since only one stimulus is involved, the relationship between the stimulus and response can be accurately quantified by experimentally manipulating the stimulus. As habituation occurs in all animals so far studied, what we learn from studies of simple organisms can provide 'models' for more complex animals. With this aim in mind a great deal of neurophysiological and biochemical investigation has been undertaken on the giant sea slug *Aplysia*. This will be discussed in a later book but here we will describe some of the behavioural responses of *Aplysia*.

Figure 6.2a shows a drawing of *Aplysia* in its undisturbed state. It is a beautiful beast with gaily coloured gills, but it has the unendearing habit of exuding a foul smelling purple ink when disturbed. This, together with its unpalatable taste, does not make it a prized gastronomic delight. It can thus move slowly over the sea bed, subsisting on a vegetarian diet and being largely ignored by predators. Its respiratory organ is a delicate gill that must be protected from damage. This could occur if a naive predator tried to eat a small portion but damage would be more likely to be sustained by another animal, such as a crab, knocking into *Aplysia*. The gill can be protected by being retracted, as is shown in Figure 6.2b. Reflex retraction of the gill can also be elicited by gentle, tactile stimulation of the siphon (Figure 6.2b). The extent of the response, i.e. the amount of gill retraction, can be measured.

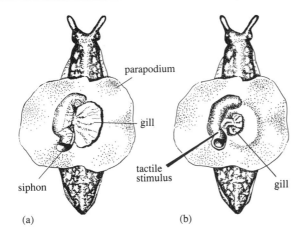

Figure 6.2   The marine slug *Aplysia*: (a) undisturbed, and (b) after a tactile stimulus is applied to the siphon, showing the gill retracted.

Figure 6.3 shows an experimental set-up for investigating the gill retraction reflex. The animal is constrained within a small aquarium and, in order to produce gill retraction, the siphon is stimulated with a jet of water (the water pik). This allows tactile stimulation to be delivered in an exact, controlled fashion, so that each stimulus is exactly the same as all the preceding stimuli. The extent of the gill retraction is measured by means of a photocell, so that as the gill retracts, more light falls on the photocell. The first response is measured and then each successive response is expressed as a percentage of the original response. Figure 6.4 (*overleaf*) shows behavioural results which are typical of *Aplysia* during habituation.

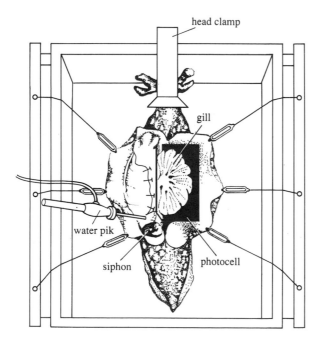

Figure 6.3   Experimental set-up for investigating the gill retraction reflex in *Aplysia*.

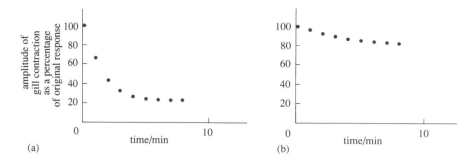

Figure 6.4 Habituation in *Aplysia*: (a) to a weak stimulus, and (b) to a strong stimulus.

☐ Compare Figure 6.4a with Figure 6.4b. What is the difference?

■ The change in response is much greater for the weak stimulus than it is for the strong stimulus, i.e. weak stimuli are habituated to more easily than strong stimuli.

☐ Can you give a functional explanation for these results?

■ Although neither of these experimental stimuli will damage the animal, in nature a strong stimulus is more likely to herald the advent of a situation which could give rise to gill damage than is a weak stimulus. Therefore, habituating quickly to strong stimuli may not be adaptive.

If the animal is allowed to rest, full recovery of the gill retraction response can be obtained. This is known as **spontaneous recovery**. After a rest of half an hour some spontaneous recovery is seen but the response of gill retraction is not quite restored in full. The intensity of the retraction may not be *quite* as strong as the original value recorded by the photocell. This is shown in Figure 6.5. Individual animals vary, but after a single training session habituation is retained for, at most, only a few hours. However, if four or more training sessions are given at 90-minute intervals then habituation is retained for up to 3 weeks.

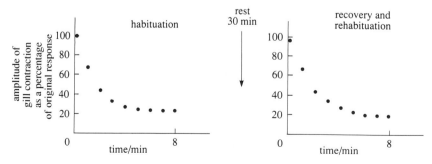

Figure 6.5 Habituation and spontaneous recovery in *Aplysia*. Note the decline in the intensity of gill retraction as the reflex is stimulated several times. A 30-minute rest period allows considerable recovery.

These behavioural responses have been assumed to be a form of learning. It has been established, using neurophysiological methods, that these responses are not simply the result of muscles becoming fatigued or of sensory receptors failing to respond (sensory adaptation). Behavioural experiments have also shown that habituation is not brought about simply by muscle fatigue or sensory adaptation.

When the same experimental arrangement is used as before, with the gill-retraction reflex being repeatedly stimulated by the water pik and habituation occurring, then a novel tactile stimulus (such as a tap with a stick) given to another part of the body, such as to the head, results in the head being withdrawn. The gill-retraction reflex is, however, also affected in that on the next presentation of the water pik the gill-retraction reflex is given in full. This is termed **dishabituation** and is shown in Figure 6.6. Dishabituation is the recovery of response strength following some other strong stimulus. This immediate recovery of full response strength without any period of rest shows that habituation is not a result of muscle fatigue or sensory adaptation.

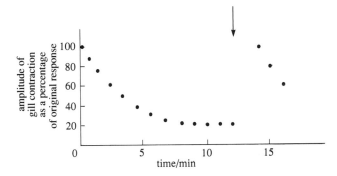

Figure 6.6 Habituation and dishabituation in *Aplysia*. Presentation of a strong tactile stimulus, e.g. to the head (the arrow), is followed immediately by dishabituation, i.e. recovery of response strength on the next presentation of the original stimulus.

The function of dishabituation is to alert the animal to a change in the environment. Habituation results in the animal not responding, or responding minimally, to a stimulus which has had no noxious consequences so far. However, the strong novel stimulus may predict danger and the animal that dishabituates will have a greater chance of survival.

Another form of non-associative learning is **sensitization**. This is an *increase* in responsiveness to a standard stimulus. For example, if you hear an unexpected sound after dark you become more alert and may respond more dramatically than you would normally to a draught moving the curtains. Like habituation, sensitization appears to be a phenomenon common to all animals.

Sensitization can be demonstrated experimentally in the common octopus (*Octopus vulgaris*). The octopus spends most of its day hidden by rocks and weeds, emerging only to attack potential prey, such as crabs, and retreating into a crevice if it sees potential danger. If it is kept in large, seawater aquaria and provided with a rocky retreat then, just as in nature, it will spend most of its time at the entrance to this retreat. If a plastic disc is placed in the tank it is likely to be ignored. However, if the octopus has experienced a small electric shock before the disc is placed in the tank, it is likely to respond to the disc by retreating into its crevice, whilst if it has recently been fed it is likely to respond to the disc by rushing out to attack it as though it were a potential prey item.

☐   What is the adaptive value of sensitization?

■   To alert the animal to a possible environmental change, so that it can respond appropriately to the change.

An environment which has become noxious to the octopus may yet harbour further danger. The octopus that shows an increase in wariness has a better chance of survival. On the other hand, if an animal that has just found food is more likely to explore, it will increase its chance of exploiting further food supplies. This assumes that these events, the presence of a predator or of food, persist over a period of time. Observations made in the field support this assumption.

☐   Is dishabituation the same thing as sensitization?

■   Dishabituation, the recovery of a previously habituated response by presentation of a strong novel stimulus, is an example of sensitization.

Responses that have not shown habituation can be increased to above normal levels by the same sensitizing stimulus that causes dishabituation. Thus, sensitization is an active process working in the opposite direction to habituation, and dishabituation is just one example of its action.

Similar kinds of behavioural responses can be recorded from other animals, including mammals (see Figure 6.7). This makes it appear realistic to study the neurophysiology and biochemistry of habituation and sensitization in *Aplysia* (as will be done in Book 4) and take the results as a basis for modelling more complex learning situations. Whilst spontaneous recovery suggests that habituation is a short term process, the fact that full recovery of the response is not obtained in the medium term argues for there being a long term mechanism involved too. Evidence for there being two stages of memory formation is presented in Chapter 4 of Book 4 in relation to the human experience. On the other hand we should perhaps show some caution in extrapolating from *Aplysia* to humans just because there are behavioural similarities.

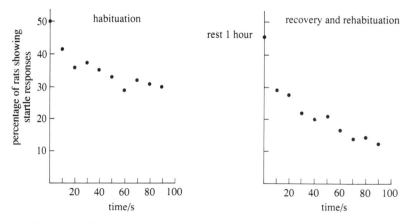

Figure 6.7 Habituation of the startle response in rats to a brief noise, presented every 10 seconds. Note that the startle response is not always given, that these results were obtained from 80 rats at each time interval, and that average response rates have been plotted.

In fact, habituation may be more complex than the preceding account indicates in that it can be context-specific. To take the example of the advertisement again, you might habituate to an advertisement on a hoarding but notice the same advertisement if it were suddenly to appear in a new place.

6.3.2   Associative learning

Associative learning involves an animal making an association between two stimuli. If you put crumbs on a bird-table after breakfast, for example, there are a number of possible associations that birds could make between the appearance of the food and related events. For example, they might learn to associate the food with the site of the bird table, the time of day, or your appearance.

Environmental changes have causes. If the cause can be identified by the birds then (because cause always precedes the effect) the consequent change (or effect) can be predicted. This would allow the animal to take appropriate action. In the example above, the bird might learn the association between you walking to the bird-table and the appearance thereafter of food. In this example, the birds have correctly associated you with the food. Remember, however, that correlation does not prove causality (Chapter 2). It is not necessary for an association to be made between the true cause and its effect for learning to occur. *Any* event which is reliably associated with the cause can serve as a predictor. Take the bird-table example again: if you always go to the dustbin immediately before putting food on the bird-table then the bird could learn the association between the sound of the dustbin being used and the appearance of food. The sound of the dustbin would create an *expectancy* of food. The bird would then fly to the vicinity of the bird-table when it heard the dustbin being used, even though the dustbin does not cause the food to appear.

In the first half of the 20th century, psychologists interested in learning investigated the conditions under which associative learning brought about behavioural modification. Two varieties of learning were described—**classical conditioning** and **instrumental conditioning**. These are essentially two different *procedures* for producing behaviour modification, and are both seen nowadays as bringing about associative learning. In the example given above, the bird makes an association between two events over which it has no direct control. This type of associative learning is brought about by classical conditioning. But animals can also make associations between two events where one event is an action that they are performing. The squirrel opening a nut learns that gnawing in a particular way results in more rapid access to the kernel. Associative learning of this kind is brought about by instrumental conditioning because the animal's own behaviour is *instrumental* in the sequence of events.

Associative learning is also involved in *passive avoidance learning*, as when an animal begins to eat something with an unpleasant taste and avoids it thereafter. The food object becomes a predictor of an unpleasant or aversive experience, i.e. the animal forms an association between the act of eating and the aversive experience that accompanies or follows it. The importance of such associative learning is obvious in relation to food selection and the learning of which potential foods are palatable and which are not. Passive avoidance learning will not be discussed further here but will be taken up again in Book 4.

6.3.3 Classical conditioning

Classical conditioning experiments use procedures that result in an animal responding to a stimulus in a novel way. For example, classical conditioning procedures can be used to modify reflex behaviour where a stimulus elicits an automatic response. In humans a puff of air to the eye will elicit the eye-blink reflex response. The puff of air is termed the **unconditional stimulus** (UCS) because it *unconditionally* (i.e. inevitably) elicits the eye-blink response and, for a similar reason, the eye-blink is termed the **unconditional response** (UCR). Under certain conditions, however, it is possible for the reflex response to be elicited by a new, and previously neutral, stimulus. This can be achieved experimentally by pairing the **neutral stimulus** (NS) with the unconditional stimulus. First, a neutral stimulus is given to a human subject; for example the sound of a musical note (a tone). The subject may respond in some way, but will probably not blink. Then, a puff of air is directed at the subject's eye; this will elicit the eye-blink reflex. Next, the subject is given a period of training; the puff of air is directed a number of times (each time is called a *trial*) at the subject's eye, but immediately before each puff of air is administered, the tone is sounded. After the training phase, the subject is tested by giving them both the UCS (the puff of air) and the NS (the tone) alone. The subject will still respond to the UCS given on its own, by blinking, but they will also respond to the tone given on its own, with no puff of air following, with a blink. Once this has happened the tone is termed the **conditional stimulus** (CS) and the elicited eye-blink is termed the **conditional response** (CR). The subject has clearly learned something and is now responding to the tone in a novel way. It has been suggested that, on hearing the tone, an *expectancy* of a puff of air is created and this elicits the eye-blink response.

☐ Could this have adaptive value?

■ Yes. If this kind of learning allows an animal to anticipate future events it can make an appropriate response more rapidly.

The Russian physiologist Ivan Pavlov (1849–1936) was the first to describe classical conditioning. He studied salivation, which is a reflex response given to the presence of food in the mouth, i.e. the unconditional stimulus of food elicits the unconditional response of salivation. His subjects were dogs and the neutral stimulus used was the ring of a bell. His experimental set-up is shown in Figure 6.8.

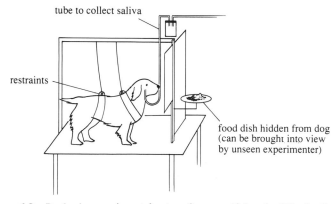

Figure 6.8 Pavlov's experimental set-up for quantifying the CR of salivation.

The dogs learned to salivate to the neutral stimulus of the bell ringing alone. Pavlov found, in addition, that this conditional stimulus could, on its own, be used to get the dog to respond to a second neutral stimulus. For example, once the bell was established as a CS it could be preceded by a light, still in the absence of food. The light was a neutral stimulus which, as you might expect, did not at first cause the dog to salivate. However, after several trials of pairing the light and the bell in the way described, the dog would salivate to the stimulus of the light alone. This was despite the fact that the light had never been associated with the presentation of food. This is known as *second-order conditioning*.

For simplicity, many experiments that use classical conditioning techniques to study learning have manipulated reflex behaviour. Classical conditioning is, however, not restricted to reflex responses.

Pavlov was aware that salivation was not the only response given by the dog to the sound of the bell.

☐ What else do you think the dog might do?

■ If it were free to move it would orient toward and approach the place where food was given. It might also lick its lips and wag its tail.

However, by restricting the dog in a harness, Pavlov concentrated upon the response of salivation. He found that one important factor in the conditioning process is the time interval between the presentation of the NS (bell) and the UCS (food in the mouth). The animal learned the association most readily when the NS preceded the UCS by about half a second.

Once a response can be reliably elicited by a conditional stimulus the effect can be long lasting. Pavlov found that a dog would still salivate at the sound of a bell even if it had not been exposed to the CS for over a year. Nevertheless it is fairly easy to disrupt the association between conditional stimulus and conditional response. The procedure is known as **extinction** and the most effective way of extinguishing a conditional response is to present the conditional stimulus repeatedly without the unconditional stimulus, i.e. ring the bell repeatedly without giving the dog any food.

Pavlov described one experiment where food was withheld for seven trials. On the first trial the dog secreted ten drops of saliva when the bell rang; by the seventh trial, only three drops were collected. However, after a 23-minute rest period the bell was rung again without food, and the animal produced six drops of saliva. Whilst this may not seem a momentous increase it is a result which can be replicated in similar conditioning experiments and the phenomenon is called *spontaneous recovery*.

☐ Where have you met this term before?

■ In the section on habituation (Section 6.3.1).

Pavlov described a further phenomenon which could occur during extinction which may also give you a feeling of 'déjà vu'. If during a series of trials where the response was being extinguished, by presentation of the bell without the food, there suddenly intruded an extraneous and unrelated stimulus (perhaps an unwary assistant bursting into the laboratory) then the response of salivation actually increased in strength on that trial.

☐ What phenomenon is this similar to?

■ Dishabituation. In fact it is called Pavlovian dishabituation.

☐ What difference can you see between extinction and habituation?

■ Extinction is the loss or reduction of a learned response, whereas habituation is the loss or reduction of an unlearned response.

Despite noticing that the dog's response to the bell was more than just salivation, in other words the response given to the bell was different from the response given to food in the mouth, Pavlov thought that, through association, the bell became a 'substitute' for the food. The notion that classical conditioning procedures lead to the formation of just a simple association between a stimulus and a response has, however, been shown to be inadequate to account for the processes underlying this kind of learning. A number of experiments have been performed which illustrate this, one of which will be described below, but first consider their general rationale.

Any stimulus which gives advance information about a change in the environment allows the animal to be better prepared. Even if that stimulus only predicts events half of the time, it will be better than nothing and the association will be worth learning. If a stimulus of greater predictive value is available, however, then it would be adaptive if *that* were the association that the animal learned. In other words the learning process should be selective about which associations are learned.

Allan Wagner and associates working in America in the mid-60s carried out a series of experiments using the eye-blink response in rabbits, which demonstrate this point (Wagner *et al.*, 1968).

In the training phase of the first experiment a light was presented together with one of two tones (T_1 or T_2). In half of the trials nothing further happened. In the other half of the trials the light preceded a puff of air which elicits the eye-blink response. On half these occasions the light was presented with T_1, on the other half T_2 was used. The conditions used in Experiment 1 are shown in Table 6.1a.

Table 6.1 Conditions used in the training phases of Wagner's experiments on rabbits.

(a)	Experiment 1	(i)	T_1	+	light	+	puff of air
		(ii)	T_1	+	light		
		(iii)	T_2	+	light	+	puff of air
		(iv)	T_2	+	light		
(b)	Experiment 2	(i)	T_1	+	light	+	puff of air
		(ii)	T_2	+	light		

T_1 and T_2 are different tones. In each experiment, the various conditions were presented an equal number of times, but they were presented in a random order.

In the test phase the rabbit was presented with just one stimulus at a time (i.e. the light, or T_1, or T_2) and its response was recorded. Results are given in Figure 6.9a.

☐ Study Figure 6.9a. To which stimulus/stimuli is a response given?

■ The light elicits a response most reliably (i.e. on about 80% of occasions) but some response is also given to each of the two tones.

In a second experiment, the training phase was different from that of the first experiment only in that for the trials where the puff of air was presented the preceding event was always the presentation of the light, L, and the tone, T_1. Thus there were only two conditions used in Experiment 2, as shown in Table 6.1b.

In the test phase the rabbit was once again presented with just one stimulus at a time and its response recorded. Results are given in Figure 6.9b.

☐ Study Figure 6.9b. To which stimulus or stimuli is a response given?

■ T_1 now elicits a response most reliably.

☐ How can this be explained?

■ T_1 always precedes the puff of air. The rabbit has learned that T_1 is a reliable predictor of the puff of air.

☐ In Experiment 2, how often is the light followed by the puff of air? (Look at Table 6.1b.)

■ The light is followed by the puff of air in 50% of the trials in Experiment 2.

☐ In Experiment 1, how often is the light followed by the puff of air? (Look at Table 6.1a.)

■ The light is followed by the puff of air in 50% of the trials in Experiment 1.

This means that in both experiments the number of pairings between the light and the puff of air is the same. Yet what the animal learns in these two experiments is different. If the number of pairings between the two events, light and puff of air, were all that were important for the learning process, then the learning would be the same in both experiments. Instead we see that in Experiment 1, the animal learns to respond to the light which acts as the best available predictor of the puff of air, i.e. the light is followed by a puff of air on half of all occasions, whereas each of the tones is followed by a puff of air on only a quarter of occasions. However, in Experiment 2, although the light has the same value as a predictor, i.e. it is followed by a puff of air on half of all occasions, the tone T_1 is a *better* predictor, i.e. it is always followed by a puff of air. The animal learns *this* association in preference to the light + puff of air association, which suggests that it is the *relative* predictive value of the stimulus that is important in the learning process.

In classical conditioning the animal is passive in that there is nothing it can do to bring about the changes in its environment. The experimenter arranges the association between stimuli or it occurs naturally in the animal's environment, and this is not altered by the animal's response. Wagner's experiments suggest, however, that the animal's learning is not a passive activity. The preceding

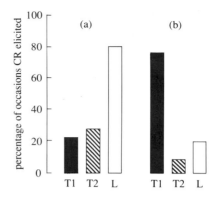

Figure 6.9 Classical conditioning of the eye-blink response in the rabbit: (a) when tone 1 (T_1) + light (L), and tone 2 (T_2) + L had each been followed by a puff of air on a random 50% of trials, and (b) when T_1 + L were always followed by a puff of air and T_2 + L were never followed by a puff of air; T_1 + L were presented randomly on 50% of the trials, T_2 + L on the other 50% of the trials.

experiments show that even when an animal is unable to control environmental events it can assimilate information about these events (or stimuli) and respond appropriately. This ability to learn the significance of an event or stimulus is the main feature shown in classical conditioning.

6.3.4 Instrumental conditioning

During classical conditioning an animal has no control over events that are changing in its environment. There are, however, a great many ways in which an animal can manipulate its environment. It should be adaptive to be able to learn the consequences of its own behaviour. Instrumental conditioning experiments also allow the processes underlying learning to be investigated.

In a particular kind of instrumental conditioning called **operant conditioning**, the experimenter sets up a situation in which an animal can behave spontaneously and where behaviour of interest can be performed repeatedly without interference from the experimenter. Behaviour patterns performed spontaneously by animals, but for which the experimenter wishes to change the likelihood of their being performed, are called **operants**. This contrasts with the behavioural responses which are *elicited* by specific stimuli during classical conditioning.

Some psychologists use the terms instrumental and operant as synonyms, but when studying the ability of an animal to learn how to find its way out of a maze the experimenter must keep replacing the animal in the maze, so maze-escape behaviour is not described as an operant although it is instrumental in effecting the animal's release. Performance generally improves in succeeding trials so instrumental conditioning is taking place.

In operant conditioning, the animal, by performing a particular behaviour pattern, brings about a change in its environment. Although operant conditioning techniques for achieving desired behavioural modifications had long been known by those who trained and worked with domestic and circus animals, it was the American psychologist, Burrhus F. Skinner who, in the 1930s, developed a sophisticated apparatus that enabled researchers to study operant conditioning more easily. This apparatus, now known as a Skinner box, is shown in Figure 6.10. It consists of a small cage which, ideally, is sound-proof. A lever (for the rat) or button (for the pigeon) protrudes through one side panel. Pressing the lever or button causes a small food pellet, or a volume of fluid, to drop automatically into a cup. As you can see from Figure 6.10 there is not a great deal of room inside a Skinner box to do anything very much, and so the animal will probably show some kind of exploratory or manipulative behaviour directed at the lever or button before too long. Eventually it will 'accidentally' press the lever or button with sufficient force to activate the food releasing mechanism. Thereafter a rat, for example, deprived of food or water will gradually learn to press the lever repeatedly to obtain the commodity of which it has been deprived. The process of learning the association between lever and food or water may take a long time, however, and to speed it up Skinner devised a procedure called **shaping**, giving food 'rewards' to the hungry rat for approximations to the desired final operant of lever pressing. Thus, in the first instance, he might reward any activity in the vicinity of the lever, then he would only reward the rat if it touched the lever. The criteria for gaining the reward became gradually stricter until finally the animal was reliably pressing the lever and obtaining food, only stopping when it was satiated.

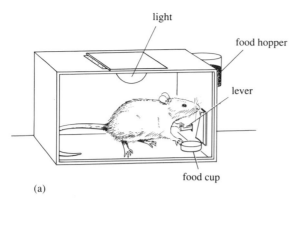

(a)

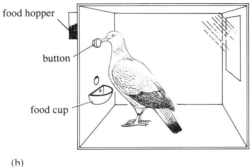

(b)

Figure 6.10   Skinner box with: (a) a rat, and (b) a pigeon.

The Skinner box has been used to study various aspects of the acquisition, maintenance and loss of this kind of learning. The process of strengthening the association between the lever pressing and the delivery of food has been termed **reinforcement**. The reward—food or water for a deprived animal—is also called the reinforcing stimulus or reinforcer. Reinforcing stimuli that increase the likelihood that the behaviour which led to them will occur again are called positive reinforcers, and the process of strengthening, the association between the operant and the positive reinforcer, is called **positive reinforcement**.

The probability that a particular operant will be shown can also be increased if the behaviour leads to the *avoidance* or *removal* of an unpleasant stimulus. Skinner called this process **negative reinforcement**. For example, a rat is placed in a cage divided into two compartments. If it stays in the compartment it is placed in, it gets an electric shock. If it moves into the other compartment within, say 5 seconds, it does not get an electric shock. By performing the operant of running from one compartment to the other, within 5 seconds, the rat *avoids* the negative reinforcement of the electric shock.

Skinner also described two conditions under which the operant is no longer performed: experimental extinction and **punishment**. Experimental extinction occurs when the reinforcer no longer follows the operant.

☐   Can you see any similarity with the extinction of a classically conditioned response?

■ In both cases the behavioural modification is lost. The conditioned response is no longer shown; the operant is no longer performed.

In both cases the loss is brought about by the breaking of an association of some kind. In classical conditioning the association between the conditional and the unconditional stimulus is broken. In operant conditioning the association between the operant and the reinforcing stimulus is removed.

If an experimental animal has experienced periods of responding but with no reinforcement, interrupted by periods with reinforcement, then the behaviour pattern becomes more resistant to extinction.

☐ Can you think of an example from every day human life where infrequent, intermittent reinforcement results in a behaviour that is particularly resistant to extinction?

■ Possibly the attraction that gambling (e.g. using fruit machines) has for some people.

More importantly, this may be a mechanism which ensures that when an animal has located a seasonal food source it does not forget about it but keeps making occasional visits to the place where it found the food. In time, when the next crop is produced, this behaviour will be reinforced.

In punishment, performing the operant results in a noxious stimulus such as an electric shock, and under these conditions, not surprisingly, the probability that the operant will be shown decreases rapidly. However, punishment can have a number of side effects: unpredictable behaviour can result and also there may be suppression of *all* the behaviour patterns that were being performed when the punishment was given, not just the behaviour pattern that was actually being punished. In other words it seems that, if the animal fails to learn the 'correct' association instantly, the punishment increases anxiety or stress and makes learning even more difficult. Stress is discussed in more detail in Book 5.

☐ What is the distinction between punishment and negative reinforcement?

■ In punishment the noxious stimulus is presented contingent upon a particular behaviour pattern (e.g. every time a rat presses a lever it gets an electric shock) so the animal learns *not* to perform that behaviour pattern. In negative reinforcement, performance of the operant *removes* the noxious stimulus or allows the animal to *avoid* it (e.g. the rat has to press the lever to turn the shock off) so the animal learns to perform that behaviour pattern.

In his experiments Skinner concentrated on positive reinforcement and this technique has been developed and used to some effect in many aspects of human behaviour, ranging from trying to encourage acceptable standards of social behaviour in children by giving positive reinforcement (say a sweet) rather than by punishment (say a smack) to sophisticated computer-based teaching machines and programmed learning texts.

Skinner also showed that it is possible for an operant to become associated with a stimulus other than the reinforcer, as long as this second stimulus is initially presented at the same time as the reinforcer. For example, if every time a reinforcer is presented to a rat in a Skinner box, a light is also switched on, then, after a while, that light acquires some of the properties of the true, or primary reinforcer. Thereafter the behaviour of pressing the lever will be performed for some time even if the light alone is presented, without the food reinforcer. A stimulus that would not normally, in itself, be reinforcing, but which acquires reinforcing properties in this way is called a secondary reinforcer.

☐ Does this process of *secondary reinforcement* remind you of any phenomenon which you encountered when reading about classical conditioning?

■ It appears to be similar to second-order conditioning.

An experiment using chimpanzees (*Pan troglodytes*) illustrates the concept of secondary reinforcement. The chimpanzees learnt to put small poker chips into a machine that provided a grape for every poker chip. Once the chimpanzees had learnt this task, they could be taught to do all sorts of other tasks just to receive the poker chips. Thus the poker chips became secondary reinforcers and, even when the grape dispensing machine was temporarily removed, chimpanzees would continue to work on their tasks to accumulate chips that they could 'spend' later. You can make your own parallels with human behaviour!

The preceding experiments and examples show that when an animal behaves in a particular way it can assimilate information about the consequences of its behaviour. In other words, in operant conditioning the animal learns the outcome of its own behaviour and modifies its subsequent behaviour accordingly.

Associative learning is so called because there is always an association between two events. In classical conditioning the association is made between a previously neutral stimulus and the unconditional stimulus (e.g. the light and the puff of air). In operant conditioning the association is between the operant and the reinforcer (e.g. pressing the lever and food delivery). Learning as a result of classical and operant conditioning appears to have many very similar features. In trying to understand the processes underlying learning, Wagner's experiments using classical conditioning techniques have enabled us to reject the idea that the formation of simple associations are an adequate explanation of this process. Similar conclusions have been arrived at from experiments using operant conditioning techniques but they are not described here. The techniques of classical and operant conditioning have allowed scientists to study some of the conditions that bring about learning but do not, on their own, allow us to say anything about underlying mechanisms in the nervous system and whether these mechanisms are the same for all types of learning. The processes involved between perceiving the stimulus and producing the behavioural modification are complex and will be discussed further in Chapter 8, as well as in Book 4. In the next section we look at more complex types of learning, such as insight and the ability to reason, and question whether these abilities could be based on the same processes as are associative learning.

6.3.5 Complex learning

There are several examples of learning that seem to be more complex than associative learning, such as imprinting and song-learning. Complex learning can also involve **insight** or **reasoning**. Insight is the ability to perceive the solution to a problem without actually working through the intervening steps and without using trial and error. Reasoning is the ability to follow through a series of abstract associations in order to arrive at a theoretical conclusion. Humans can solve some problems using insight and reasoning and it is suggested that they are not unique in possessing these abilities. However, the following experiments show that it is difficult to get scientists to agree about this subject.

Chimpanzees, supplied with boxes and/or bamboo poles that can be fitted together, will quickly obtain food that has been placed beyond their reach, e.g. outside the cage or suspended above them, by fitting together the bamboo poles and using them as rakes or by stacking up boxes to get closer to the food. It has been claimed that this shows the chimpanzees possess insight because the chimps go through only a short period of trial and error before discovering how to reach the food—they were able to reach it almost straight away.

However, in other experiments, laboratory-reared chimpanzees not previously exposed to sticks were given pairs of sticks that could be fitted together. Nineteen out of twenty adults completed the novel task of fitting the sticks together in less than 5 minutes. Similarly, given boxes in their cages, chimps not previously exposed to boxes will stack them and climb on them. So these activities, performed in the absence of food and which might be described as play or exploration could form part of these animal's normal behavioural repertoire. As such they can be used to obtain food on occasions when food is present, but out of reach.

Thus it is clearly important to know about the learning history of the animal before carrying out this kind of study. The crucial question then is whether the animal really is showing insight, or whether the problem is solved by trial and error. The chimpanzees can show a considerable variety of behaviours before they get to the food and some of these seem to be irrelevant to the task in hand, suggesting that trial and error may be more important than insight.

Individual chimpanzees may solve problems in different ways, and they have the ability to learn from others. This latter ability is widespread among animals. It is often called **imitation learning** and is particularly useful if the animal lives in a group (see Chapter 10 for more examples).

It is not always clear what exactly are the necessary conditions (or experiences) for complex learning to occur. For instance, an attempt was made to reduce the numbers of the common crow in the USA using poisoned bait, but it failed because after a few deaths the rest of the flock avoided the bait. Are the crows demonstrating insight, or is some simpler process at work based upon their earlier learning experiences?

In their normal environment individual members of a species will be subject to very similar early experiences and so the opportunities for learning will be very similar. Nevertheless learning is a property of the individual and each individual's life experience is unique. Thus when studying the current learning ability of an individual within a population it must be remembered that that ability may itself depend on some form of prior learning. It is quite difficult to be certain about the opportunities for learning that have occurred in any individual's life-time and this makes some of the experiments which purport to demonstrate insight and the ability to reason open to other interpretations.

Some primates do seem able to reason, but you can judge for yourself from this description of an experiment using rhesus monkeys (*Macaca mulatta*) which was designed to avoid the complications of prior learning. The monkeys were given a choice of two objects. When object A was lifted it revealed a peanut, whereas lifting object B revealed nothing. The monkey had to learn to make the association of lifting A to obtain food. The animal learned this and was then given the next problem: object C covered a peanut, object D covered nothing. Once the monkey learned the association between C and food a further problem was set involving two new objects, and so on. These problems are called two-choice discrimination problems and the monkeys were set, literally, hundreds of them.

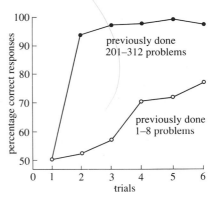

Figure 6.11   Performance of rhesus monkeys over the first six trials of a series of two-choice discrimination problems. The lower curve represents average performance on the first eight problems, the upper represents performance after previous training on at least 200 problems.

☐   Now look at Figure 6.11. How well did the monkeys do on the first trial, on average?

■   There is a choice of two objects so there is a 50% chance of making a correct choice, and 50% of the monkeys got the choice correct.

☐   What happens on the second trial?

■   Monkeys that had solved this sort of problem up to eight times before did not choose the correct object more often than would be expected by chance (i.e. about 50% chose correctly). Those that had already solved 200 or more such problems, however, chose the correct object in over 90% of cases.

The monkeys seem to have learned something about the type of problem they are being required to solve. What they have now learned has been described as a 'win–stay, lose–shift strategy'. They are still forming an association between object and food but after solving a number of similar problems they have been able to extract some feature or features of the total experimental set-up in such a way that after only one trial with the new object an *expectancy* is created. If they remove the object that was covering the food item on the first trial after training on 200 similar problems then they will remove that same object on the second trial, expecting that once again food will be present. In other words they stay with their original successful choice ('win–stay'). But if they chose the wrong object on the first trial and find no food then, on the second trial, they will shift their choice to the other object in the expectation that it will contain the food ('lose–shift'). The monkey is described as having 'learned to learn'. What appears to have changed over this problem-solving series is the way in which the stimuli presented during the experiment are being processed by the animal.

More difficult reasoning tasks can involve *second-order relationships*. In these tasks, if an animal is given a two-choice discrimination problem, as just described, and is then given a second stimulus which is the 'clue' to the correct response then the information given by the second (or conditional) stimulus (or cue) can be used in solving the problem. Suppose we again have objects A and B which are covering a peanut or empty space. But the food is not always under A. It is under A when a buzzer sounds but under B when a bell rings. Many animals can learn to use the conditional stimuli to determine where the food is.

The situation can be altered so that the conditional cues share some feature of the objects between which the animal must choose. Say that A and B are both similar bowls except that A is red and B is green. Now the conditional cue could be a red light or a green light. If the red light shines and the red bowl contains the food it is called a *matching to sample problem*. (The alternative is an *oddity problem*, where the red light shines and the green bowl is the correct response.) The question is whether the animal can learn the abstract concept of choosing 'same'. If they can, then having learned that a red light indicates that the red bowl contains the food, and given a choice between a square bowl and a round bowl they would choose the square bowl when a square panel was illuminated and the round bowl when a round panel was illuminated. Pigeons cannot do this but experiments have shown that rhesus monkeys and dolphins can, even when the actual stimulus is changed on every trial. The processes underlying this ability might be somewhat different from those involved in simple associative learning, but they are not necessarily so. The ability to reason and show insight will be considered again in Chapter 8.

6.4   Constraints on learning

6.4.1   The evidence for constraints on learning

If the processes underlying learning are the same in all species then it seems reasonable to study the rat in the laboratory in order to elucidate these processes. However, in attempting to do this, many psychologists lost sight of what animals do in their natural environment and it came as something of a surprise and disappointment to discover that not all species learn all tasks equally well. One of Skinner's students, Keller Breland, was so impressed by Skinnerian techniques that he gave up his career in psychology and with his wife Marion set up a small business to provide animals for the world of entertainment. They had many successes but were frustrated in some of their attempts, so much so that they wrote a paper (Breland and Breland, 1961) entitled 'The misbehaviour of organisms' (a title which parodies one of Skinner's influential works 'The behaviour of organisms'). They had worked with 38 different animal species and had had great success but were struck by the *types* of failure they encountered. For example, domestic pigs were trained to drop coins into a piggy bank. They were reinforced for *carrying* the coin in their mouths but, even after long periods of training, they persisted in 'rooting' these coins along the ground. Although the pigs could perform the required carrying behaviour and this behaviour was reinforced, the Brelands could not get the pigs to retain this operant. They describe this as 'a clear and utter failure of conditioning theory' because their training in psychology had led them to understand that the *law of equipotentiality* operates, i.e. that all animals can learn all tasks equally well (within the limits of their sensory and motor capabilities).

The pigs' rooting behaviour is a typical component of the animal's normal food gathering behaviour and the Brelands suggested that, where an operant is close to an 'instinctive behavioural pattern' (i.e. species-specific behaviour pattern), it will drift towards the more natural species-specific behaviour pattern. This they called **instinctive drift**. Thus, through instinctive drift, the form of the operant becomes adulterated.

Some behaviour patterns such as face washing, scratching with the hind leg, and scent marking in hamsters, which are not normally associated with food, are particularly resistant to conditioning to obtain a food reward. On the other hand, there are operants that would in the animal's natural habitat relate, in a way that is biologically significant, to the reinforcer being employed and these are much easier to condition. That this may be so is demonstrated rather dramatically by watching pigeons in Skinner boxes peck at buttons. When food is the reinforcer the pigeons peck with their beaks open, making energetic stabs at the button as they do when feeding. If water is the reinforcer the beak is closed, contact with the button is maintained for longer and swallowing movements may be seen, just as though the pigeon is drinking (see Figure 6.12).

There seem to be similar constraints on avoidance behaviour. A rat quickly learns to escape from a 'dangerous' location A to a 'safe' location B in response to receiving an electric shock at A. If a rat is placed in a large box in location A, and then, after 5 seconds, it is given a shock whilst still in A, it will quickly learn when it is subsequently replaced in A to move from A before the 5 seconds have elapsed. As a result it will not receive further shocks.

☐   Has this learning been achieved by negative reinforcement or punishment?

■   Negative reinforcement. If the rat leaves A it will avoid the shock.

However, rats will learn with great difficulty to shuttle back and forth between A and B to avoid a shock on either side of a barrier. What they have to learn here is that A is sometimes safe, i.e. when B is dangerous, but at other times A is dangerous when B is safe. The experiment requires the rat to return immediately to a 'safe' location that was 'dangerous' a few seconds ago. It may be that the first experimental situation resembles situations in the natural environment in which the rat has to avoid noxious stimuli, whereas the second experimental situation is unlike any natural situation. The animal is said to be *prepared* to learn certain things and *contraprepared* to learn others.

This realization that there were constraints on learning led some psychologists to adopt a more 'natural history' approach to their studies. They began to investigate animals' learning abilities, based on the supposition that the animals would be particularly good at tasks which in their natural habitat would have biological significance. They might therefore be able to recruit capacities adapted to such tasks and show learning under conditions which had previously been thought unsuitable. For example, it had been thought that for associations to be learned the events in question must be closely associated in time. Pavlov had found optimal conditioning when the conditional stimulus preceded the unconditional stimulus by half a second. The following experiment, the essential details of which are summarized in Figure 6.13 (*overleaf*), shows that this is not always necessary.

Two groups of thirsty rats were allowed to drink from a drinking spout. Group 1 rats drank water with a novel flavour (saccharin). Group 2 rats drank 'bright noisy water', i.e. the water was unflavoured but as the rat's tongue made contact with the drinking spout it activated a flashing light and a bell.

Each group of rats was further sub-divided into Groups A and B for the next phase of the experiment. Group A rats were given an electrical shock immediately after drinking. Group B rats were left alone for some hours then X-irradiated. This induces a feeling of nausea.

In subsequent tests Group 2A rats ('bright noisy water' followed by shock) would not drink water from the drinking spout, but Group 1A (novel flavour followed by

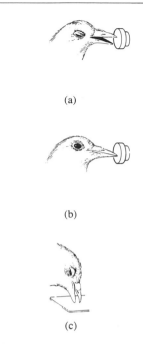

(a)

(b)

(c)

Figure 6.12 Pigeons pecking in Skinner boxes. (a) Button-pecking in a pigeon with food as a reinforcer. Note the closed eye and open beak, typical of pigeon feeding behaviour. (b) Button-pecking with water as a reinforcer. Note the open eye and closed beak, typical of pigeon drinking. (c) An attempt was made to train a pigeon to press a lever with its foot. Instead it pecks the lever. This is an example of instinctive drift.

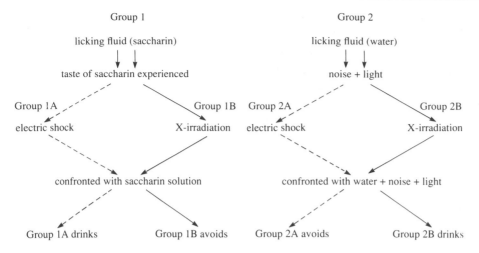

Summary of the 'bright–noisy water' experiment

		OUTCOME	
		illness (X-irradiation)	electric shock
STIMULI	taste (saccharin)	1B avoidance	1A no avoidance
	visual and auditory (light and sound)	2B no avoidance	2A avoidance

Figure 6.13 Outline of procedure for and results of the 'bright noisy water' experiment.

shock) would drink saccharin in another location. In other words rats can learn to associate the location of drinking water where lights flash and bells ring with a subsequent shock and will avoid the 'bright noisy water' location in future. However, drinking flavoured water just before receiving a shock does not result in the rat associating the particular flavour of the water with the shock. In contrast Group 1B rats (novel flavoured water followed by X rays) would not drink the flavoured water again, even though the nausea occurred hours after drinking. Group 2B drank water from the same location again—the 'bright noisy water' did not become associated with the sickness. In other words, rats learn 'taste leads to nausea' and 'location leads to shock' associations, but not 'location leads to nausea' or 'taste leads to shock' associations. This effect is most readily shown when the taste is a novel one. Rats exposed to plain water followed by sickness do not subsequently avoid the plain water.

The predisposition to learn certain things and not others makes good biological sense here. The rat normally identifies food by its flavour. However there is normally a relatively long delay between the ingestion of harmful substances and ill effects being experienced, so you can appreciate that there is considerable adaptive value in being able to make the association between the flavour and the subsequent illness even when there is a long delay between the two events. Similarly flashing lights and bells ringing, whilst being a remarkably uncommon occurrence in the wild, has about it the makings of an appropriate prelude to some nasty physical event such as the appearance of a predator. (The parallel here of the emergency services' sirens and flashing coloured lights is obvious!) So it is not surprising that the rat makes an association between the visual and auditory stimuli and the subsequent shock, and that it is contraprepared to associate the taste of drinking a particular fluid with the shock.

In fact, rats and some other species have been shown to be able to learn about food by making very indirect associations. Rats can detect, by smell, the type of food that has been recently eaten by another member of their social group. When subsequently given a choice of flavoured food they choose to eat the same flavour as their group-mate in preference to other foods. Other studies indicate that they will avoid food which has been eaten by another individual if that individual is subsequently ill, just like the crows mentioned earlier.

6.4.2   Implications of the 'constraints on learning' evidence for traditional learning theories

Earlier in this chapter it was stated that psychologists have tried to find general laws of learning, often in the belief that learning is a unitary process, common to all organisms. Those who are critical of this idea have cited the 'constraints on learning' evidence in refuting it, and suggest instead that there are several different mechanisms or processes of learning. At the time of writing the question remains open.

Ethologists have always related behavioural studies to field observations and therefore, as mentioned earlier, their preoccupation is with species-typical behaviour patterns and their functions. Thus ethological studies of learning have focused on phenomena such as imprinting and the sensitive periods in song learning, and the 'constraints on learning' evidence is not at odds with their findings. More recently there have been cross-species studies of learning, investigating differences in behavioural responses and also in brain structure. For example, within the tit (*Parus*) family there are species that are storers and species that are non-storers of food. Observations in the wild have shown that food storers store items in hundreds of different places and that they can find these items again after periods of many days. Some experiments have suggested that learning abilities between these two types of birds may not differ greatly if recall is tested relatively soon (i.e. 30 minutes) after the learning opportunity has occurred but that memory may be more persistent in food storers (such as coal tits, *P. ater* and marsh tits, *P. palustris*) than it is in non-food storers (such as the blue tit, *P. caeruleus* and the great tit, *P. major*). In Chapters 7 and 8 it will be argued that this kind of ability depends on the animal being able to learn about spatial relationships within its environment.

The 'constraints on learning' evidence does, more seriously, challenge the assumption of the law of equipotentiality, i.e. that any individual can learn anything. It seems more likely that there are genetic differences between species in what they can learn and how easily they can learn it. However, some psychologists disagree with this view and suggest that failure to learn, such as rats failing to learn to avoid the 'bright noisy water', occurs only because of prior learning which 'gets in the way' of this new situation. In other words the animal comes to the learning situation with prior expectancies. For example, rats drink plain water so frequently and the outcome is usually beneficial so it is to be expected that the association of nausea with drinking plain water will be blocked by the prior learning that water has beneficial consequences. The argument becomes much more tortuous, however, when examining the reasons for the rat failing to associate the novel saccharin flavour with the shock, and it need not be gone into here.

To try to explain all the constraints on learning evidence in terms of prior learning and subsequent blocking does seem doomed to failure. It seems unreasonable to deny the possibility that some species start life with a predisposition to learn certain associations, for example, a flavour–nausea association has been shown to exist in rat pups; further examples are imprinting and the learning of bird song.

So is there a place for attempting to construct a unitary 'Theory of Learning' and associated laws? The diversity of learning and the constraints which arise even when looking at relatively simple tasks performed in the controlled laboratory environment might suggest that the ability to perform these tasks depends on different physiological mechanisms. *Hydra*, a small aquatic animal with a very simple nervous system, habituates to certain stimuli but shows no evidence of any associative learning. Classical conditioning of limb reflexes has been demonstrated in animals in which the spinal cord is intact but the brain is absent, whereas the brain seems to be necessary for operant conditioning to be effective. Primates have the most complex and extensive nervous systems in terms of brain structures and they also show the most complex learning abilities. Whether all this indicates that different physiological mechanisms are necessary for learning processes *per se* is not clear—for example, it may be that the brain is necessary for operant conditioning because, without it, normal sensory and motor processes will be unable to provide sufficient information to be associated by the learning process. It could be that the processes which underlie all learning are the same. Animals do, in other respects, show more uniformity at the biochemical level than at the behavioural level. You will read more about investigations into the molecular basis of learning in Book 4.

Summary of Chapter 6

Learning is the acquisition of new knowledge or a new skill. It is a widespread phenomenon but it is often difficult to prove that it has occurred and to be certain of the necessary conditions for it to occur. Learning is inextricably linked with memory and recall so an animal that apparently fails to learn may, in fact, be demonstrating a memory loss or a recall problem.

Learning occurs most readily when it has consequences for the animal's fitness. Being able to learn gives flexibility to an animal's behavioural repertoire.

Learning has been studied by many scientific disciplines. The psychological approach has tried to find general mechanisms underlying learning. Psychologists have distinguished three different categories of learning. First, there is non-associative learning, e.g. habituation and sensitization, where the animal alters its behaviour after it is repeatedly exposed to a single stimulus or event. Second, there is associative learning where an animal shows evidence of making the association between two events or stimuli that have a particular relationship to one another; classical and instrumental (particularly operant) conditioning are two techniques that have been used to investigate associative learning. Finally, complex learning is a term used for all other learning abilities, such as imprinting, song learning, insight and the ability to reason.

The law of equipotentiality states that all animals can learn all tasks equally well. This is challenged by evidence that there are constraints on learning and has led to many psychologists adopting more of a 'natural history' approach to learning studies.

There is still uncertainty as to whether all the behavioural examples given in this chapter can be classified as 'learning', that is, whether it is a unitary process, with general laws covering its acquisition.

Objectives for Chapter 6

When you have completed this chapter you should be able to:

6.1   Define and use, or recognize definitions and applications of each of the terms printed in **bold** in the text.

6.2   Give examples of learning and explain why it is so difficult to give an all embracing definition of learning. (*Questions 6.1 and 6.2*)

6.3   State the adaptive value of learning. (*Question 6.3*)

6.4   Describe and distinguish between the ethological and psychological approaches to the study of learning.

6.5   Explain, giving examples, what is meant by habituation and sensitization. (*Questions 6.4 and 6.5*)

6.6   Explain, giving examples, what is meant by associative learning and distinguish between classical and operant conditioning. (*Question 6.4*)

6.7   Give examples of complex learning.

6.8   Cite evidence for the existence of constraints on learning abilities and discuss whether this undermines: (a) the theory that there is a common process or mechanism underlying all learning, and (b) the law of equipotentiality. (*Question 6.3*)

Questions for Chapter 6

Question 6.1 (*Objective 6.2*)
A male puppy squats to urinate, like a female. An adult male cocks his leg. How could you investigate whether this change in behaviour was a result of sexual maturation, e.g. the presence of adult levels of sex hormones, rather than of learning.

Question 6.2 (*Objective 6.2*)
How would you investigate whether rats can learn the location of food, even though, at the time of learning, they are neither hungry nor do they ingest the food?

Question 6.3 (*Objectives 6.3 and 6.8*)

Whereas rats rapidly form associations between the taste and smell of food and subsequent ill effects, some species of bird can more easily be conditioned to form an aversion to visual stimuli. Predict how rats and birds might discriminate between possible foods in their natural habitat.

Question 6.4 (*Objectives 6.5 and 6.6*)

Is sensitization the same process as learning brought about by classical conditioning, as described in the Pavlovian situation of the salivary response to a bell?

Question 6.5 (*Objective 6.5*)

Compare Figures 6.5 and 6.7; both show habituation and recovery and rehabituation, but Figure 6.5 shows results from an invertebrate (*Aplysia*) and Figure 6.7 shows results from a vertebrate (rat).

(a) Suggest a reason why the time interval between the presentation of successive stimuli might differ in these two experiments.

(b) What significance can be attached to the fact that, after the rest periods chosen (30 minutes and 60 minutes, respectively), recovery of the original response is not quite complete.

References

Breland, K. and Breland, M. (1961) The misbehaviour of organisms, *American Psychologist*, **16**, pp. 661–664.

Wagner, A. R., Logan, F. A., Haberlandt, K. and Price, T. (1968) Stimulus selection in animal discrimination learning, *Journal of Experimental Psychology*, **76**, pp. 171–180.

Further reading

Halliday, T. R. and Slater, P. J. B. (eds) (1983) *Animal Behaviour, Vol. 3: Genes, Development and Learning*, Blackwell Scientific Publications.

CHAPTER 7
MOTIVATION

7.1   Introduction: what is motivation?

Consider a laboratory rat with free access to food and water. It is housed in a cage and exposed to a daily rhythm of lighting of 12 hours dark followed by 12 hours light. It eats and drinks predominantly in the dark. Sometimes a rat eats nothing in the light and spends most of its time sleeping. Behaviour in the constant presence of food and water is therefore distinctly not constant.

If the same rat is deprived of food during a period when it would normally eat and then is given food, it will avidly eat it. It then refrains from eating for a time although food is still present. Something happens inside the animal that plays a part in the sequence of eating, and not eating. We could say that it is first motivated to eat and then, as a result of ingestion, is unmotivated. We assume that the change in feeding motivation is to do with the state of its body. First, it is depleted of energy and then repleted. Thus feeding motivation can vary with changes internal to the animal.

The experiment is repeated but after the rat shows signs of losing its feeding motivation we replace its usual food with food that the animal has not ingested for some time. Typically, the rat immediately ingests the new food; it is 'motivated' to eat it. So motivation also concerns external events. Some foods are more attractive, more motivating than others.

Motivation concerns the *variability* of an individual's behaviour. Sometimes animals eat and sometimes they drink. At other times they ignore food and water. How do we explain the variability of behaviour? Suppose that we deprive a rat of food for a period when it would normally be eating and then allow it to eat. During deprivation, changes occur within its body; for example, reserves of energy will be used up. When we allow food again, the amount ingested depends upon the length of deprivation; feeding motivation increases during deprivation. Why not simply say that the longer the animal is without food, the more it will eat when food is available? Or why not measure the change in its body, e.g. the lowering of the blood glucose level, and associate this with behaviour? How does the term 'motivation' help?

One main reason why we need the term is that our knowledge of the important events within the body that underlie the variability of behaviour is still inadequate. In the case of feeding, we suspect that it has something to do with blood glucose but we are far from sure.

For the motivation underlying feeding, drinking and sex, as well as other behaviour patterns, there is an intense research effort to find the physiological basis in the brain and elsewhere in the body. However, at best we can only identify some of the important factors. We do not know all of them or their exact mode of interaction. To say that motivation is increasing with deprivation and this increase can often be reflected in more than one measure of behaviour is sometimes the best we can offer.

Even if we did know exactly what is the important physiological variable that changes during deprivation, we would still need the term 'motivation'. Changes in the amount eaten can be induced by means other than deprivation, e.g. increasing the attractiveness of food. Associating the food with nausea (see Chapter 6) for example, will decrease the amount ingested. Motivation refers to a proposed internal variable that can be influenced by several different procedures.

The motivational changes that take place during deprivation are not only reflected in the amount subsequently eaten. For example, normally the longer the period of deprivation is, the more likely a rat is to press the lever in a Skinner box (described in Chapter 6) to obtain food and the less discriminating it is concerning what it eats.

A similar logic applies to sexual motivation. In species such as rats and hamsters, one way of measuring male sexual motivation is the number of mounts, intromissions (penile insertions) and ejaculations that the male shows in a mating session. Figure 7.1 indicates that as the time from the previous mating increases, so the male's responsiveness increases. We do not know exactly what happens in the body during recovery but we refer to the animal at, say, 10 days, as having higher sexual motivation than at 1 day. Figure 7.2 shows the behaviour of rats that were castrated and then given androgens (a type of sex hormone secreted mainly by the testes). Motivation, as measured by number of copulations, is increased by replacing the androgens lost as a result of castration. Both androgens and deprivation could therefore be said to increase sexual motivation.

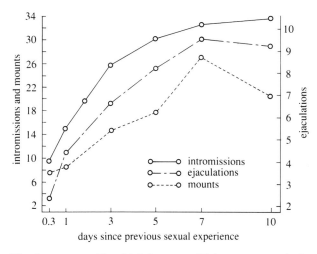

Figure 7.1 The frequency with which hamsters (*Cricetus auratus*) showed different aspects of sexual response at various times after recovery from their previous sexual experience.

By motivation, we imply something happening within the body of the animal, most probably within its brain. Exactly how we should envisage its basis is still not clear. However, to simplify, you might regard motivation as being the activity of certain nerve cells. In the case of feeding motivation, this activity would depend in part upon the energy balance of the animal. The rat's feeding motivation changes over the course of a meal, i.e. the activity of a particular set of nerve cells changes. Feeding motivation plays a part in bringing about feeding

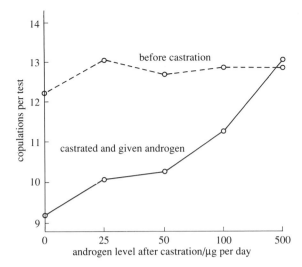

Figure 7.2   Sexual response (measured by number of copulations) of male rats castrated and then given various amounts of replacement androgen.

activity and when motivation is lowered, the animal will no longer pursue this activity. In this case, feeding motivation underlies feeding. There will be other motivations, such as those underlying drinking and sexual behaviour, which will also have some level and which will change in value as circumstances change.

Psychologists envisage that each motivation is part of what is termed a **motivational system**. For example, the feeding motivational system is composed of feeding motivation, the behaviour involved in the ingestion of food, the physiology involved in the passage of food through the gut and the absorption of nutrients, and the detectors somewhere in the body that monitor the animal's energy state. One interest of psychologists is in understanding the details of the various motivational systems, such as feeding, drinking and sex.

Changing motivation can be invoked to account for variation in behaviour even when external stimuli are constant. However, injury, maturation and learning also produce variation in behaviour. So how do we distinguish motivational changes from these? In practice, most scientists have little difficulty doing so, but inordinate difficulty in giving a water-tight definition of the distinction. However, we must make an attempt at such demarcation. Changes in behaviour that are attributable to learning arise from the animal's *experience* of the environment and involve assimilation of information about such experiences (Chapters 5 and 6).

A sexually immature animal does not respond to sexual stimuli in the same way as an adult. An animal that has been injured may also alter its behaviour. If, for example, an insect loses a leg, it adopts a new pattern of leg movements when walking. Both these situations involve a change in behaviour in a given environment, but we would not consider them to be motivational changes. One factor that distinguishes the process of maturation from motivation is that maturation is generally not reversible. An animal does not fluctuate between being adult and being immature but it does fluctuate between being hungry and not hungry. The effect of injuries is often a permanent one. Even where it is not, injury concerns pathology, things going wrong, whereas motivation concerns intact processes. In such a context, some would define motivation as 'that class of *reversible* internal processes responsible for variations in response to external stimuli'.

Food does not invariably elicit feeding and we say that whether it does so depends upon the animal's motivation. Using motivation to refer to the variability of behaviour, what sort of system might be described as *not* being motivational? For healthy humans, a tap on the knee *invariably* elicits a knee-jerk response and a bright light invariably elicits constriction of the eye's pupil. The effectiveness of such stimuli does not fluctuate. It would be quite inappropriate to postulate that people are, or are not, motivated to straighten their knees when they are tapped. These processes are described as reflexes. Their failure to appear is not an indication of some natural underlying change but warns of serious disease or malfunction.

In addition to variable behaviour, motivation is also associated with behaviour being directed towards a goal, as the following example shows. A rat has been deprived of water and is placed in the start-box of a maze. On arriving at a particular part of the maze, it finds a bowl of very salty water. It tastes this, which is, of course, inappropriate for its physiological needs, and does not ingest it. Only water, or a weak solution, would be of any use in restoring its dehydrated state. In future ,when tested in the maze following water deprivation, it avoids the part of the maze containing the salty water. Later the rat is maintained on water *ad libitum* (meaning always freely available) but made deficient in salt. It is then returned to the maze. Typically, it rapidly negotiates its way through the maze towards the salty water and ingests it. Some psychologists speak of the rat as having the goal or purpose of obtaining salt. The location of salt was learned on the first encounter and the memory remained latent until the animal's physiological condition made it the goal of behaviour. In these terms, the motivation induced by sodium deficiency was responsible for selecting the location of salt in the maze as the goal to be pursued, as opposed to any other goal.

In addition to motivation being associated with variability in behaviour and its goal-directed nature, it concerns *decisions*. Animals do not just feed or drink; they must decide when to switch from one activity to another. The word 'decision' does not necessarily imply any conscious reasoning process analogous to human decision making but rather that, at some point, behaviour has to change from one pattern to another. Whatever the exact nature of the process, the nervous system must have the facility to switch between one behaviour and another. Consider an ethologist describing the sequence of activity of an animal in the wild that he or she is observing. First, the animal is feeding, then it switches to drinking and then in response to an alarm call it dives for cover. The ethologist is interested in the fitness benefits the animal derives from switching between behaviours. Clearly, the animal's chances of survival and reproduction are increased by the facility to switch rapidly and appropriately between different activities. At times of emergency, all activities other than escape must be temporarily suspended. The ethologist might speak of competition between the various motivations for expression in behaviour. For example, the strength of the motivation to drink might become higher than that to feed and the animal would then switch from feeding to drinking; the appearance of a predator, however, would instantly introduce a powerful defence motivation which would suppress the influence of such motivations as those of feeding and drinking. This assumes that different motivations can be measured in terms of a *common currency*, which enables the strengths of different motivations to be compared. So, in the context of switching between behaviour patterns, motivation implies the existence of a series of states that can change in strength relative to each other and which can play a role in organizing behaviour in the sense of promoting some activities but inhibiting others. Seen in this way an animal that, for instance, stops feeding cannot necessarily be assumed to have a zero level of feeding motivation. Rather, a stronger motivation might have appeared and so the animal changes what it is doing.

Finally, in some cases motivation is able to *selectively potentiate* certain species-typical behaviours, i.e. a particular motivation leads to a specific behaviour pattern. For example, a frightened rat having no escape possibilities typically freezes. All rats tend to freeze in much the same way. They do not have to learn how to do it by trial and error, unlike, say, lever-pressing in a Skinner box. Freezing is potentiated by fear. A sexually aroused female rat will display a pattern of body arching, called lordosis, in the presence of an aroused male, as shown in Figure 7.3. This pattern, characteristic of rats, facilitates the male's mounting attempts. We would say that sexual motivation in the female serves to potentiate or prime the key species-typical and motivation-specific motor act of lordosis. When the female's hormonal state changes and her sexual motivation is no longer present, males are not pursued and an advance by a male fails to elicit the mating posture. Motivation is therefore involved in the stopping as well as the starting of behaviour.

Figure 7.3   Mating in rats. Note the female's lordosis posture.

These different aspects of behaviour for which the concept of motivation is employed will be discussed throughout this chapter. Some researchers place more emphasis upon one use than another, but it will be argued that they are all appropriate for a comprehensive picture. Thus motivation typically fluctuates in strength and it causes appropriate goals to be selected and pursued. When relatively strong it potentiates the behaviour with which it is associated and promotes that behaviour in competition with other possible behaviour patterns by exerting inhibition upon them. Finally, it can potentiate certain species-typical behaviour patterns.

7.2 Changing responsiveness—internal factors

7.2.1 Introduction

Changes in reaction to food or water that occur as a result of changes in the internal condition of the body are intimately connected with the physiological functioning of the body. In a similar way, low body temperature motivates an animal to perform behaviour patterns that lead to the gain of heat or reduction of heat loss. As discussed earlier, variability of responsiveness to a constant external stimulus can also be observed for sexual motivation. A male rat having performed a series of ejaculations is less responsive to a receptive female than one that has not recently mated. As the period of deprivation from sexual contact increases, some animals show a fall in discrimination, meaning that a wider range of stimuli are adequate to arouse mating.

In Section 7.1, we looked at four aspects of behaviour for which the concept of motivation is used: (1) changing responsiveness to a constant stimulus, (2) goal-directedness, (3) switching between behaviour patterns, and (4) potentiation of species-typical behaviours. Most motivational systems fit these criteria. However, there are some systems, concerned with defensive behaviour, that might be termed 'motivational' by criteria 2 to 4, but not by criterion 1; in these systems responsiveness to a given stimulus fluctuates little if at all. Suppose an animal has recently defended itself against something such as a predator by fighting or fleeing. It would be maladaptive if the tendency to show such behaviour were subsequently lowered and only recovered with time. A second predator might appear immediately after the first. Indeed, the evidence is that, in general, performing such defensive behaviour does not lower motivation. This takes us to the heart of considering the *function* that motivational systems serve and the relationship between this function and the nature of the causal mechanisms involved in each.

☐ What is the function served by feeding and drinking? Is it possible to relate this to the causal mechanisms that underlie behaviour?

■ Both feeding and drinking serve the regulation of the physiological condition of the body, i.e. an internal state. One might expect a deficiency of food and water to motivate feeding and drinking, respectively, and, of course, this is the case.

☐ What is the function served by sex? Is it possible to relate this to the causal mechanisms that underlie behaviour?

■ Sex does not serve regulation of physiological variables apart from fertilization and reproduction. One might therefore expect sexual motivation to be relatively easy to arouse at times when fertilization chances are maximal. In the case of the females of many species, sexual motivation does indeed fluctuate with which stage of the reproductive cycle she is in (Chapter 9).

☐ What is the function served by defensive behaviour? Is it possible to relate this to the causal mechanisms that underlie behaviour?

■ This serves to protect the animal from damage or death and to conserve resources. As noted already, one might expect it to be permanently primed for action.

The internal state of humans can be changed and they can be asked how their reaction to various stimuli changes. We are all familiar with such changes; food generally tastes better the longer you have been deprived. Of course, in parallel, the motivation to eat increases as deprivation increases. We can quantify scientifically such subjective reports. In a series of now-famous experiments, Michel Cabanac (1979) asked human subjects to rate on a numerical scale the pleasure of stimuli and observe how the pleasure rating changes as a function of internal state. Figure 7.4 shows one such result. The pleasure rating of a temperature stimulus varies from positive, through indifference, to negative, as a function of core body temperature.

☐   In what way do the results shown in Figure 7.4 relate to the function served by body temperature regulation?

■   The human body functions optimally at a core temperature of 37 °C. In a state of hypothermia (low body temperature), it is advantageous to favour a warm environment, and to be motivated to seek one, so as to elevate body temperature. When the body is hyperthermic (high body temperature) it is advantageous to move to a cool environment so as to lower body temperature.

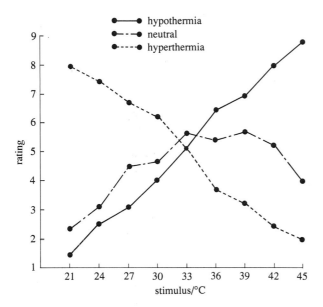

Figure 7.4   The subjective rating of a temperature stimulus by human subjects (vertical axis) as a function of the temperature of the stimulus to which their body is exposed (horizontal axis). Ratings: 1 = very disagreeable, 3 = disagreeable, 5 = indifferent, 7 = agreeable and 9 = very agreeable. Subjects were tested when at a normal ('neutral') body temperature, when at a hyperthermic (above normal) body temperature, and at a hypothermic (below normal) body temperature. Note the shift in pleasure rating in the two situations where body temperature deviates from the norm (hypothermia and hyperthermia).

7.2.2 Homeostasis

The previous section gave an example of the regulation of an internal state at an optimum value. Normally, the body temperature of mammals and birds is maintained close to a particular value, which in humans is 37 °C. If temperature departs far from this, either up or down, the body functions less well. A substantial departure leads to death. Therefore physiological mechanisms have evolved that serve to regulate body temperature close to the desired value.

☐   Can you name some of these mechanisms and what they achieve?

■   Sweating and panting are induced by hyperthermia and serve to lower body temperature. Shivering is induced by hypothermia and serves to raise body temperature.

Other mechanisms maintain other internal states. If for example, body-fluid levels rise, urine formation is increased until the excess fluid level is corrected. Conversely, if body fluids are depleted, the rate of urine formation drops. The maintenance of a variable within narrow limits is called **homeostasis**, a word literally meaning 'close to standing still'. Homeostatic mechanisms are of great importance in the normal functioning of animals. Any change in body state beyond a certain limit triggers a physiological process which serves to restore the body to the desired state. Departures from normal limits also affect motivation, however. A low level of body fluids not only promotes a reduction in urine formation, but also increases drinking motivation.

☐ How do departures from normal body temperature affect behaviour?

■ Hyperthermia for example promotes behaviour which tends to move the animal into a cooler environment, while hypothermia promotes behaviour which tends to move the animal into a warmer environment.

For rats in the laboratory, hypothermia can motivate, for example, the operant response of pressing a lever in a Skinner box for the reward of a unit of hot air. In humans, it motivates the response of pressing the heater button in a British Rail waiting room. In some animals including humans, huddling occurs, though there is a possibility that the motivation here can be somewhat mixed!

The possession of both intrinsic physiological and motivational homeostatic mechanisms (involving the external world) is of obvious adaptive value. They mean that an animal can venture into and survive in relatively hostile territory, for example, extremes of heat and cold.

Departures from homeostatic norms trigger motivation which in turn promotes a change in physiology and/or behaviour that tends to correct the departure and restore the *status quo*. Thus homeostatic systems exhibit what is termed **negative feedback** (see Figure 7.5). A feedback system is one in which a factor A influences B, but B also influences A. Thus, for example, body-fluid level influences drinking but drinking also influences body-fluid level. The greater the difference between the actual body-fluid level and the 'desired' body-fluid level, the greater the effect on drinking and so the quicker the normal body-fluid level will be restored. Feedback is 'negative' because drinking *increases* when body fluid levels *decrease* and vice versa. If drinking increased when body fluid levels increased, feedback would be 'positive'. **Positive feedback** thus leads to a greater and greater departure from the original level. In everyday language, positive feedback is often called a 'vicious circle' (see Section 7.6).

☐ How does temperature regulation illustrate negative feedback?

■ A *rise* of body temperature causes physiological (e.g. sweating) and behavioural (e.g. moving to a cooler environment) responses. These responses *lower* body temperature. Conversely, a *drop* in temperature triggers responses that cause body temperature to *rise*. These actions continue until the optimum temperature (37 °C in humans) is reached.

This can be compared with the regulation of room temperature by a thermostat. The occupant sets a temperature on the dial, say 20 °C. The actual temperature is compared with this so-called 'set-point' temperature. If the actual temperature

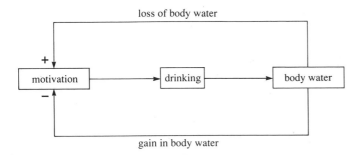

Figure 7.5 Illustration of the principle of negative feedback. The body loses water by various routes (e.g. urination). This has an excitatory effect upon drinking motivation (indicated by +). Drinking motivation causes the animal to drink. The body fluids gain water by drinking, restoring levels to normal and so lowering drinking motivation (indicated by −).

falls below the set-point, the heater is switched on. Heat is generated, the room temperature is increased until actual and set-point temperatures are equal and then the heating is switched off. In some home temperature regulation systems a cooling system is switched on when the temperature rises above the set-point level. This analogy is useful in understanding motivation, but like all analogies it must not be pushed too far. There are some fundamental differences between such a system and biological negative feedback systems, as will become apparent later. The principle of homeostasis can be most clearly illustrated by looking at the physiological details of an individual system and this characterizes the approach of physiological psychology.

Traditionally, physiological psychology attempts to understand motivation by understanding the underlying physiological components and the homeostatic mechanisms involved. For example, it is known that animals drink when depleted of water and that drinking reduces motivation. Therefore an understanding of the physiology of body fluids would be expected to give insight into the motivation underlying drinking.

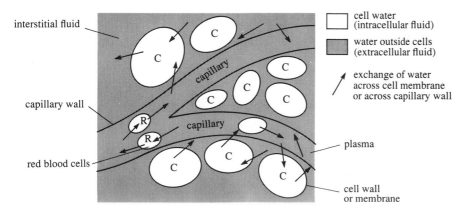

Figure 7.6 Cells of the body (C) showing the membrane that surrounds them and the fluid medium (the interstitial fluid) that occupies the spaces between them. Also shown are some red blood cells (R) in a capillary (a small blood vessel).

The body of a rat, like all mammals including humans, is composed of about 70% water. Some of this is within the cells of the body and some of it outside the cells (see Figure 7.6). The water within the cells (e.g. brain, kidney, skin cells) is known as the **intracellular** water, or **intracellular** compartment. This term does not imply a continuous single volume of water as in a bucket. Rather it simply means that, for our purposes of explanation, the water in the cells can be treated as if it were a single body of water. All of the water outside the cells constitutes the **extracellular** water or **extracellular** compartment. It is made up of the water in the blood plus that in the spaces between the cells.

Loss of water from either compartment is sufficient to motivate drinking. Removal of blood results in an extracellular deficit. Injection of concentrated salt solution into the bloodstream creates an intracellular deficit by causing water to move out of the cells and into the blood, so that the salt solution is diluted. Somewhere in the body, there must be detectors that monitor the state of each compartment. It is unlikely that the whole volume of the compartment could in any way be measured. Rather the evidence is that 'samples' are taken. A cell or a group of cells is monitored for their volume and, in effect, they serve as a representative of the whole cellular compartment. In vertebrates, these cells are located in an area of the brain called the hypothalamus. As far as the extracellular fluid is concerned, there appears to be a receptor that detects the amount of stretch in a particular blood vessel. When blood volume falls, the detector signals this to the brain and drinking is motivated.

So much for what switches drinking on. What serves to terminate it? Could it be correction of the dehydrated states within the intracellular and extracellular compartments? Consider for a moment the time factors involved, as represented in Figure 7.7. The animal drinks, water passes down the oesophagus (gullet) and into the stomach, where some is absorbed. The rest then passes into the intestine from where it is absorbed through the wall of the intestine into the blood. Only when it is in the blood is it described as being part of the body fluids 'proper'. (For these purposes the gut is described as being outside the body, since it is basically a tube running from mouth to anus and substances within the tube have not yet passed across the body wall.) Some of this water then passes from the blood into the cells.

Clearly, correction of dehydration in the extracellular and, particularly, intracellular compartments takes a significant time. Yet an animal such as a rat or dog will rapidly ingest an amount of water commensurate with the size of its deficit and then stop. At the time of stopping, most of the water will still be in the stomach and intestine. Little will have been absorbed across the intestine wall. Clearly a factor other than correction of dehydration of intracellular and extracellular compartments serves to terminate drinking. There is evidence for two inhibitory influences on drinking: water in the stomach and water passing through the mouth. Each of these represents a transient factor. Water moves out of the stomach and the memory of water passing the mouth decays with time. As they do so, the state of dehydration is corrected (see Figure 7.8).

Although negative feedback is central to understanding feeding and drinking, not every instance of such behaviour can be explained in these terms. There are also processes that sometimes can act to anticipate departures from the norm and so pre-empt them, a property termed **feedforward**. Drinking provides a good

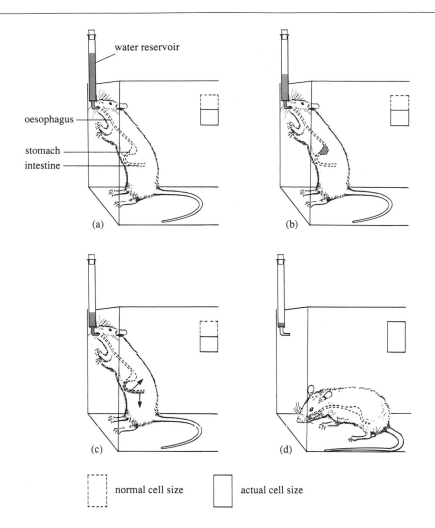

water reservoir

oesophagus

stomach

intestine

(a)

(b)

(c)

(d)

☐ (broken) normal cell size   ☐ actual cell size

Figure 7.7 Sequence of the location of water. The rectangle represents the state of a sample cell. The broken rectangle is the cell volume that is maintained when the rat has water *ad libitum*. (a) Dehydrated rat starts to drink, (b) water fills the stomach, (c) water is moving into the intestine and being absorbed into the blood. Note that there is little correction of the dehydrated cell yet. (d) The rat stops drinking. A while later water has left the gut and cell is restored to normal.

example of this. After eating dry food, water tends to be pulled from the blood into the gut of an animal (Figure 7.9, *overleaf*), which tends to dehydrate the blood. When they eat, however, rats drink substantial quantities of water. One might suppose either that they do so because of a dry mouth or because of the pull of water into the gut. In fact, neither is a necessary condition. Drinking occurs too rapidly after feeding to be explained by the movement of water across the gut. Therefore, it seems that the drinking is *in anticipation of* the consequences of ingesting food rather than in response to them. This is feedforward rather than feedback and might well imply learning. On the basis of past experience it could be that the animal has learned an association between eating and subsequent dehydration. It then drinks on the basis of this association.

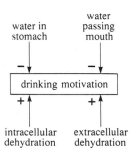

Figure 7.8 Factors contributing to drinking motivation. + represents excitation (increase in motivation). − represents inhibition (decrease in motivation).

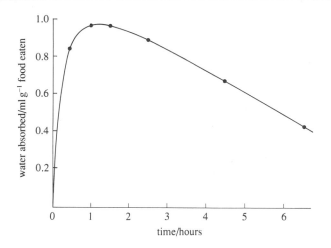

Figure 7.9 Water moving from the blood into the gut of a rat in response to the appearance of food (standard laboratory diet) in the gut.

☐ What could be the functional significance of drinking in anticipation of dehydration rather than in response to it?

■ By drinking rapidly after a meal, the animal manages to pre-empt dehydration and thereby keep body fluids relatively undisturbed. Were the system to work by negative feedback alone, the animal would eat, water would move into the gut and drinking would be aroused by the fall of body fluids. The disadvantages of relying on this control would be that body fluids would be disturbed and the gut would get very full.

Suppose an animal such as a rat or a pigeon is placed in a hot environment. Water loss from the body increases and drinking increases.

☐ What would constitute negative feedback control of drinking under these conditions?

■ The loss of water from the body would promote drinking which would then correct the loss.

Such a loss of water is indeed sufficient to motivate increased drinking but there is evidence to show that, in addition to negative feedback control, under such conditions there is also a feedforward control over drinking.

☐ What would constitute feedforward control?

■ Drinking in response to the high temperature itself and before dehydration occurs.

Experiments have shown that rats and pigeons do indeed show such anticipatory drinking. By measuring the loss of water from the body in the absence of drinking water, it can be shown that drinking occurs too rapidly on exposure to the warm environment to be explained by such loss.

7.2.3 Non-homeostatic systems

Sometimes a dichotomy is drawn between two classes of motivational system. Some involve the regulation of a physiological variable. These include feeding, drinking and temperature regulation. In each case, if regulation breaks down, then ultimately the shift from normal in the physiological variable causes death. Systems of the other class do not serve the function of regulating a physiological variable and they are sometimes termed *non-homeostatic*. Motivations in this class include sexual motivation. Lack of opportunity for sexual behaviour does not cause a shift in a physiological variable that poses a threat to the body. It is sometimes claimed that this motivation is more strongly tied to external, rather than internal, stimuli, though later it will be necessary to qualify this view.

Sexual motivation is determined in part by sex hormones. In the female, such hormones as testosterone, oestrogen and progesterone (Chapter 8) fluctuate regularly over a certain time period called an oestrous cycle, or menstrual cycle for primates (Figure 7.10a). Sexual motivation also fluctuates during the oestrous or menstrual cycle. A female rhesus monkey (*Macaca mulatta*) was placed in a Skinner box and had to press a lever 250 times in order to get access to a male from which she was separated by a barrier. Figure 7.10b shows that the time it took the female to perform 250 responses varied with the stage of her menstrual cycle. The less time she takes to perform 250 lever presses, the higher is her sexual motivation. Motivation to gain access to the male was highest at around the time of ovulation.

☐   Study Figure 7.10. What could be the hormonal basis of the female's change in motivation?

■   The blood concentrations of the hormones oestrogen and testosterone both peak just before the time of maximum sexual motivation. This could thus contribute to the level of sexual motivation.

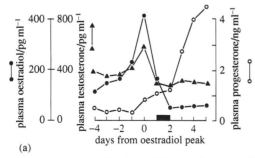

(a)

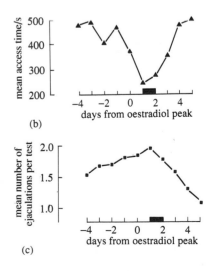

(b)

(c)

Figure 7.10   Changes in behaviour of male and female rhesus monkeys as a function of stage of the female's menstrual cycle. (a) Changes in blood levels of oestradiol (a type of oestrogen), testosterone and progesterone over a part of the oestrous cycle. (b) Time taken for female to perform 250 lever presses in order to gain access to a male. (c) Number of ejaculations exhibited by males when placed with female for fixed time period. Time of ovulation is indicated by the black bar.

In support of this interpretation, further experiments have shown that injection of these hormones into female rhesus monkeys has the effect of increasing their sexual motivation.

☐ What is the functional significance of the fluctuation in female motivation shown in Figure 7.10?

■ Sexual motivation is at a maximum at the time when fertilization is possible.

Note that in Figure 7.10c the average number of ejaculations by a male monkey placed with a female for a fixed time period was at a maximum at the time of ovulation. Thus the male–female motivational interaction is such as to maximize fertilization chances.

The hormone testosterone also makes a contribution to male sexual motivation, making it more likely that a conspecific female will stimulate the male to mate. The influence of testosterone is deduced from experiments in which the testes are removed, causing the amount of testosterone secreted by the animal to be drastically reduced (see Figure 7.2). In rats, for example, the level of male sexual activity falls within a few weeks of castration.

What lowers motivation immediately following mating? It might surprise you, but this is a little researched area. At one time a model of male sexual motivation based upon the pressure of seminal fluids in the testes was proposed. Ejaculation was thought to lower pressure and so motivation. However, male rats that are caused to ejaculate by artificial means do not show lowered motivation. This idea also leaves the existence of sexual motivation in females unexplained! The drop in motivation after mating is not caused by a lowering of sex hormone levels: these do not decline post-mating and, even if they were to, their effects on the central nervous system are not that rapid. In the male, it has been argued that some kind of change in the central nervous system follows normal ejaculation(s). This would then underlie the lowering of motivation. Possibly ejaculation causes an inhibitory neuron to be activated and its inhibition dissipates with time.

In conclusion, no identifiable physiological variable (in the sense of homeostasis) is regulated by sexual behaviour. However, the motivation of sexual behaviour *is* comparable to that of eating and drinking in that it is a function of the internal state of the animal. Finally, we should not overstate the differences between homeostatic and non-homeostatic motivational systems. Both classes of motivation depend upon external (see next section) and internal factors. Further similarities will be described later.

7.3   External factors in motivation

So far this chapter has concentrated on the role of internal factors in determining motivation. This section looks at external factors. It is important to realise, however, that motivation is best understood as the outcome of a complex interaction between internal and external factors. In some cases, understanding of one set of factors is better than that of the other. For example, the external factors relevant to exploratory behaviour are well understood, but relatively little is known about the internal factors involved.

Exploration is an example of behaviour that does not, as far as is known, serve to regulate a particular physiological variable. When an animal such as a rat or cat is placed in a new environment it will spend considerable time wandering around and exploring. Even when the environment is thoroughly familiar, such animals still spend time periodically exploring it. One of the motivating factors is a change in the environment. When something novel is introduced the disparity between the new sensory information and the old prompts exploration.

☐   What might be the functional significance of exploration?

■   Exploration allows an animal to assimilate information about its environment. Imagine an animal migrating into a new environment. It can locate potential escape routes for use in an emergency. It can locate sources of food. Later it can revise the information in the light of any changes in the environment.

The external environment is also of crucial importance in the case of feeding. As was noted earlier, even for a constant internal environment, the strength of feeding motivation can change as a result of a change in the external environment. A rat that appears satiated on one food will often avidly ingest a new food, another example of motivation being a function of both external and internal factors. Consider Figure 7.11. A standard laboratory diet was given one of four arbitrary flavours, A, B, C and D. One flavour might be, for example, lemon. The figure shows the amount ingested in a 2-hour feeding session when the rat was given access to each of the four flavours in turn, e.g. test day 1 when it received D, B, A, then C, compared with test day 2 when it received C throughout the 2 hours.

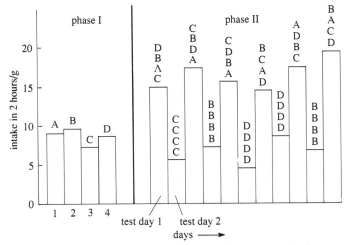

Figure 7.11   The role of variety in diet. In phase 1, on four successive days over a 2-hour period each day, four different diets are given. In phase II, during each successive daily 2-hour feeding session, either the same diet is available throughout (e.g. CCCC) or the diet is changed each 30 minutes (e.g. DBAC).

☐   What is the effect of variety in the diet?

■   Variety tends to increase the total amount eaten relative to when a single diet is available throughout.

☐   What might be the functional significance of such behaviour in the wild?

■   It ensures that, given a variety of available foods, a rat tends to eat some of each. A balanced diet has to include various vitamins and amino acids for example. These might be distributed throughout various different available foodstuffs.

It would be wrong to ask whether an animal is eating because of *either* external *or* internal factors, or which are more important in feeding. When an animal is eating it will always be because of some combination of influences arising from internal and external factors. For example, no matter how much novelty is introduced in the diet or how attractive the food is made, food intake will be constrained by the negative feedback arising from ingested food.

A similar effect to that of Figure 7.11 is shown for some species when a male apparently sexually satiated with one female is exposed to a novel female. Sexual behaviour is sometimes strongly rearoused, a phenomenon termed the **Coolidge effect**.

An interesting question for all motivational systems is how internal and external factors act in combination to determine motivation and how changes in either factor influence behaviour. This question is investigated in the following experiment.

A Siamese fighting fish (*Betta splendens*) was trained to swim through a tunnel to obtain a reward: a variation on the theme of the Skinner box. For some fish the reward was food and for others it was the opportunity to perform their aggressive display (normally performed in front of another male) to a mirror. The fish were trained on a schedule of one reward for each response made. Later the 'costs' were increased so that, for example, the fish needed to swim through the tunnel twice to obtain one reward, a response/reward ratio of 2. The fish were finally stretched to a response/reward ratio of 6. Figure 7.12 shows the result of the experiment in terms of the behaviour of the fish as a function of the response/reward ratio.

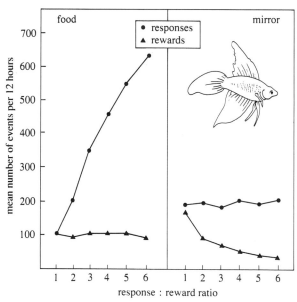

Figure 7.12 Number of responses made and rewards earned by Siamese fighting fish as a function of response/reward ratio imposed. *Left*: when food was the reward. *Right*: when the opportunity to display in a mirror was the reward.

☐ Study Figure 7.12. In what way does the fish's behaviour in response to increasing cost differ when the reinforcer is food compared with when it is a mirror?

■ When food is the reward, as the response/reward ratio is increased so the fish performs more responses to compensate. As a result, the number of rewards obtained in a period of time remains the same. When the mirror is the reward, making the costs greater is not associated with increased rate of responding by the fish. Consequently, the number of rewards obtained falls.

In this situation, performance of the aggressive display is more sensitive to changes in the external environment than is feeding. For feeding, if fewer rewards are obtained, compensatory action maintains reward input nearly constant. Some ethologists draw an analogy here with demand characteristics in human society. As Figure 7.13 shows, when the price of fish increases, consumption comes down. This is described by economists as an *elastic* relationship between price and demand. By contrast, coffee shows a relatively inelastic relationship. Consumption falls only slightly with increases in price. By this analogy, in the operant situation, increasing the response/reward ratio is analogous to increasing the cost that the animal has to pay to obtain the reward.

☐ By this analogy, do feeding and performing the aggressive display exhibit elastic or inelastic cost–demand characteristics?

■ Feeding is inelastic and performing the aggressive display is elastic.

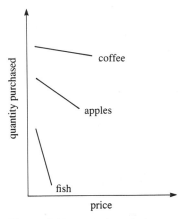

Figure 7.13 Relationship between quantity purchased and price of three household items.

One might cautiously extrapolate to the natural environment and suggest that when food supply is low, the time that the fish spends on food-seeking activities will increase to compensate. A behaviour having a more elastic cost–demand characteristic would cease to be performed at such a time.

As these examples show, each motivation arises from a combination of internal factors, such as energy, fluid or hormone level and external factors, such as available food or the nature of the operant task set the animal. For some systems, e.g. exploration, our understanding of external factors is better than for internal factors. The sensitivity of motivation to variations in external factors may be greater for some systems (e.g. aggressive behaviour) than for others (e.g. feeding). Such things as the availability of food and its quality play a crucial role in determining the strength of motivation. In addition, stimuli which in the past have been associated with such things as food, a mate or an intruder can acquire a motivational strength as a result of this association. This is an example of conditioning, discussed in Chapter 6, and it forms the topic of the next section.

7.4 The effect of learning on motivational state

Just to remind you, the procedure of classical conditioning as studied by Pavlov consisted of pairing a neutral stimulus (e.g. a bell, a light) with a biologically significant event, a UCS (e.g. food, electric shock). Following such pairing, the neutral stimulus acquires a potency that it did not possess before, and is now termed a *conditional stimulus* (CS). In modern terminology, classical conditioning is described as a relationship between one event E_1 (e.g. bell) and a subsequent event E_2 (e.g. food). The formation of a conditioned association is represented as $E_1 \rightarrow E_2$, meaning the presentation of E_1 creates an expectation of E_2. There are two sorts of associations involving external events: (1) associations between neutral events and biologically important events, and (2) associations between biologically significant events and the consequences that follow from experience of them. The next two sections look at two motivational systems to see how an understanding of learned associations helps to interpret what is going on.

7.4.1 Feeding

Feeding is a good example of where stimuli (conditional stimuli) that have been associated in the past with food (unconditional stimulus) can increase motivation. H. P. Weingarten (1984) presented a tone (T_1) together with food to rats that had been deprived of food. Later the rats were allowed food *ad libitum*, but when the tone was presented, food was eaten. Thus a stimulus that in the past had been associated with food presentation can increase feeding motivation.

☐ What procedure is necessary to make sure it was specifically the pairing of T_1 with food that was responsible for increasing feeding motivation?

■ It is necessary to present another tone at times other than when food is presented, and to test later whether this tone has the capacity to elicit feeding.

In fact, Weingarten used two tones, T_1 and T_2. For half the rats, T_1 was paired with food and T_2 was presented at random times. For the other half, T_2 was paired with food and T_1 was presented at random times. This eliminates the possibility that something intrinsic to a particular tone is able to evoke feeding rather than its pairing with food presentation. Weingarten then compared the amount ingested by satiated rats immediately following food-associated tones with that ingested following tones that had been presented at random. This procedure is summarized in Figure 7.14. The tone associated with food (but not the other tone) was able to evoke considerable feeding in 'satiated' rats.

Group 1	Group 2
RATS HUNGRY tone and food presented simultaneously	RATS HUNGRY tone and food presented independently at random times
RATS SATIATED tone ↓ feeding	RATS SATIATED tone ↓ no feeding

Figure 7.14 Summary of Weingarten's experiment.

Animals form associations between stimuli that predict when food will arrive and they also form associations between ingesting food and the consequences of ingestion. As you saw in Chapter 6, if a characteristic taste is followed by nausea, an animal can form a taste → nausea predictive association. When food of this flavour is offered in the future, the animal tends to reject it. The food might have been palatable before the experience with nausea. Therefore a model of feeding needs to take into account (1) energy state, (2) quality of available food, and (3) associations previously formed with the food. One possible interpretation is that, in the case of a taste $\rightarrow$ nausea association, a memory of the association is stored, so that presentation of the flavour revives the memory and this serves to lower motivation.

Experiments have shown clearly that the association between ingestion and its consequences can be positive as well as negative, as the following example shows. Rats are unable to detect the presence of vitamins in food but can utilize a learning process to obtain needed vitamins. If a laboratory rat is first made to suffer a vitamin deficiency by being exposed to a diet deficient in a certain vitamin, it will actually stop eating this food. If the rat is then exposed to a novel food, it will ingest it. If it happens to contain the needed vitamin, then, as a consequence of its ingestion, some time later the rat will recover from the deficiency to some extent. In future, the flavour of the novel food will be favoured by the rat.

☐   The rat would be said to have formed an $E_1 \rightarrow E_2$ association. What are E_1 and E_2?

■ E_1 is the taste of the novel food and E_2 is the recovery from the vitamin deficiency.

Such a process could play a part in diet selection by wild rats. There are a number of essential vitamins and amino acids that they need in their diet. However, the rat is not able to recognize each of these. If it finds itself deficient in an essential component, the familiar flavours will lose much of their motivational value, intake will fall and the rat's energy state will be low. A novel food will have a relatively high motivational value and will be ingested. If it proves to contain the missing dietary component it will be favoured in the future. If not, the rat will refrain from ingesting more of it.

7.4.2 Aggression

K. L. Hollis (1984) reported some observations on territorial defence in a fish, the blue gourami (*Trichogaster trichopterus*). Male fish were presented with a rival male the appearance of which was either signalled by a conditional stimulus (a red light) or unsignalled. Residents presented with a signalled rival were found to be much better prepared for the attack than those presented with an unsignalled fish, as measured by success in the ensuing fight. Hollis argues that the adaptive significance of this is clear: '...the best defense is a good offence, a strategy obtained through Pavlovian conditioning'. The warning gives the animal time to prepare. In motivation terms, the warning signal increases attack motivation.

Aggressive behaviour is also influenced by learned associations with its consequences. In some species, such as mice, the tendency to attack a new opponent increases in an individual that has had frequent experience of fighting successfully with opponents. Conversely, individuals that frequently lose fights become less likely to attack a new opponent and become more likely to show fear responses. Familiarity with an opponent can also change the level of aggression shown towards it. Many birds that establish territories in the breeding season show less and less overt aggression towards their immediate neighbours over the course of the season. Thus the same individual—'the bird next door'—changes from being a stimulus that elicits a high level of aggression to one that elicits little. Should a novel bird arrive in the vicinity, however, it is vigorously attacked, indicating that there has been no change in the underlying potential for aggression.

7.5 Goal-direction

So far the emphasis has been upon motivation as a factor in the decision to, for example, feed or not feed. This decision was shown to be influenced by both internal and external factors. However, the animal will not always be in the presence of the object appropriate for the motivation. Some animals might simply be stimulated into activity by a depletion of energy and thereby happen upon food. However, imagine a bird setting out at sunrise and flying to its usual feeding patch. It follows a particular route in order to get to this patch avoiding obstacles. One can only imagine that a combination of (a) its energy state, and (b) its *memory* of where food is located plays a role in this.

In addition to considering motivation as a determinant of whether or not to behave in a certain way, motivation is also associated with giving behaviour *direction*. For example, this might involve a location of food as an appropriate object for the motivation. In these terms, motivational systems are responsible for selecting goals to be pursued and organizing the animal's movement towards the goal. The example of salt depletion was used earlier to introduce this topic, where the motivation induced by salt depletion led to the selection of a site where concentrated salt solution was located as the goal to be sought.

A further aspect of motivation when viewed in these terms is the *flexibility* of behaviour. To understand this, consider a rat negotiating its way through a maze to obtain a food reward. Suppose, after the rat has learned the task, the maze is flooded with water. Typically, the rat will then swim through it to reach the goal-box. What is common between the two situations is not the mechanics of movement; clearly rather different movements by the legs are involved in the two cases. Nor are the stimuli exactly the same in the two cases—contact with the floor is not possible in the flooded maze. What is in common is the goal reached. The animal is able to recruit a variety of different responses to reach the same goal. Motivation therefore does not necessarily activate a fixed response. Rather motivation activates *behaviour towards a goal* that the animal pursues by whatever means are appropriate in the particular circumstances. In the case of the salt-depleted rat, the motivation induced by this state would select the goal to be reached: 'get to the location of salt'. The term 'goal' is used rather appropriately to describe such behaviour in a variety of different contexts; it is said to be *goal-directed* or *purposive*. It is as if the animal has a purpose in getting to the goal.

Take an animal engaging in defence. It could be said that the goal ('end-point') of this behaviour is to place a distance between the defender and the attacker. To achieve it, a number of different strategies can be employed; the animal can change its behaviour according to prevailing circumstances. When one strategy is succeeding, it will persist with this. When it starts to fail, a different strategy will be adopted. For example, attack, associated with aggressive motivation, might be the first strategy but, if this fails, defence motivation is likely to increase and to generate an escape strategy. A cornered rat will switch from escape to attack or to passive submission.

Defence provides a clear example of where conditioning can play a role in goal-directed behaviour. In a typical experiment, a rat is exposed to a few pairings of a tone followed by a mild shock to the feet. The reaction to the shock is one of jumping or flinching. The tone is then presented on its own. In response to the tone, if the environment affords the opportunity, the rat will escape from the situation and thereby avoid the shock altogether. For example, it might jump onto a safe perch. If there is no escape possible, the rat will typically freeze (i.e. remain immobile). So motivation does not just play a role in generating behaviour that is a fixed, stereotyped response to a given stimulus. Rather, behaviour can be modified according to prevailing circumstances. In the natural habitat, an animal such as a rat which is prey to a number of other species would be well served by efficient processes of defence involving conditioning. When the opportunity for escape is present, it might prove best to take it. If there is no opportunity for escape, freezing might prove to be the best strategy.

Why is such behaviour described as 'goal-directed'? There is the modification in behaviour that can occur according to circumstances. For a given stimulus, a novel response can appear. For example, a rat might be trained to associate a tone and a shock. Then a safe perch is introduced into the cage. Next time the tone is sounded the rat will probably jump onto the perch. It could be said that the animal is selecting a goal that is appropriate to the motivation and the opportunity afforded by the environment. Another aspect of the goal-directed or anticipatory nature of motivation is shown in a series of experiments on feeding, described next.

Two groups of rats were trained in Skinner boxes to press a lever to obtain food pellets of a distinct flavour (Dickinson, 1985). Later, in a context outside the Skinner box, the particular flavour was *devalued* for one group (Group 1) by inducing nausea after the rats had ingested it (see Chapter 6). Then, when they were fully recovered, both groups were tested in the Skinner box under extinction conditions (i.e. no food was delivered when the lever was pressed—see Chapter 6). It was found that Group 1, having experienced nausea, pressed much less before giving up than rats in Group 2. It was argued that as a result of the experience of nausea the rat learns a taste → nausea association. Activity in the Skinner box is directed towards the goal of gaining food. When the goal is devalued the rat works less vigorously to attain it. It could be said that the rat is expecting food in the Skinner box and when this expectation is associated with food having a negative association the motivation to get food is lower. Figure 7.15 shows a summary of this.

Group 1	Group 2
trained to press lever to earn pellets	trained to press lever to earn pellets
flavour paired with nausea	exposed to nausea and flavour but not at any associated time
tested in Skinner box under extinction conditions PRESSES RELATIVELY LITTLE	tested in Skinner box under extinction conditions PRESSES RELATIVELY FREQUENTLY

Figure 7.15   Summary of Dickinson's experiment.

The behaviour in the Skinner box cannot be explained in terms of the animal's reaction to the actual food in question since under extinction conditions there is no food present. Neither can it be explained in terms of an aversion developed towards the lever, since the only aversive association was made outside the Skinner box. Rather, Dickinson argues that the rat that is lever-pressing has a purpose to its activity, to gain food. When the object of this purposive activity is devalued then this will be reflected in the activity. More will be said on this subject in the next chapter.

7.6   Decision making

So far the discussion has concentrated on looking at individual systems in isolation, e.g. feeding and aggression. Traditionally, psychologists have pursued this approach to the subject. Ethologists have shown more interest in how an animal distributes its time between various activities. One activity, e.g. feeding, is usually incompatible with another, e.g. mating, and therefore the animal must make 'decisions' as to what to do. Engaged in one activity, it needs to decide if and when to do something else. Ethologists are interested in the causal mechanisms underlying such switching but are also concerned with its functional significance. This theoretical perspective is rooted in the assumption that an animal should behave in such a way as to maximize its chances of survival and its reproductive success, i.e. its fitness (see Chapter 4). A number of examples can be given to show this aspect of motivation and its implications for the topics discussed earlier.

Consider the situation shown in Figure 7.16: the loss of weight of a domestic hen while she is incubating her eggs. The hen loses weight because she eats only briefly each day during this period.

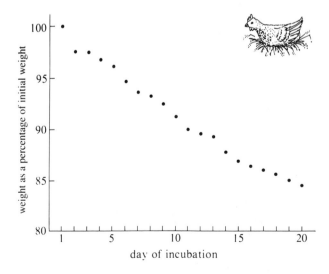

Figure 7.16   The change in weight of a hen during incubation of her eggs. Despite the fact that food is available only 50 cm from her nest, the hen does not take in sufficient food to maintain her body weight.

☐   What *adaptive* explanation might be given for her behaviour?

■   To leave her eggs often or for long periods would expose them to the risks of chilling or being eaten by predators.

For the hen there appears to be a trade-off in the relative costs and benefits of risking harm coming to the eggs and maintaining body weight. It appears that in terms of natural selection, the greater benefit lies in ensuring her reproductive potential than in maintaining her weight. Feeding might be said to show more elasticity than incubation (Section 7.3). This example illustrates two general points. First, whether the hen feeds or not cannot be explained simply by considering feeding motivation. Rather it needs to be understood in terms of this system and its *interaction* with incubation. It appears that incubating eggs is biologically 'more important' than feeding and thus takes precedence over it. Secondly, in proposing that the hen selects the strategy that is biologically most appropriate, it is not suggested that she makes a conscious calculation of the costs and benefits of different strategies and arrives at this one. Rather, it is argued that evolutionary processes have selected for those strategies that are most appropriate. Hens adopting other strategies are at a disadvantage and hence the alleles biasing them towards such strategies have prospered less well. At the causal level of explanation, it is proposed that the incubation motivation system exerts *inhibition* on the feeding motivation system.

Consider another example. A lizard in the desert has a high feeding motivation and to obtain food needs to venture into the sun. However, in so doing, it runs the risk of hyperthermia. Consider now each of the two systems of body energy/feeding and body temperature regulation in isolation. Energy state motivates the lizard to seek food on the surface. If the environment were not hostile, this system would produce a food intake commensurate with the deficiency in the body. However, in the absence of a strong feeding motivation, temperature state would be such as to cause the lizard to bury into the deeper and cooler sand. Clearly, both motivational systems cannot gain full expression in behaviour. A compromise is reached; the animal obtains some food and then buries itself into the sand. The compromise means that neither physiological state, energy or temperature, is compromised too far. On the other hand neither state is held at the optimum. In the language of set-points, neither physiological variable can be maintained at its set-point but deviations in each are restrained.

What factors play a role in the decision to switch between different activities? This is a controversial question in ethology but some of the factors can be identified. If an animal is simultaneously food-deprived and water-deprived, switching between feeding and drinking will depend to some extent on the relative strengths of the states of deprivation. If food-deprivation is much greater than water-deprivation, the animal will spend a relatively long time on feeding before switching to drinking. From functional considerations, switching between activities is expected to depend also upon the ease with which the animal is able to switch. For example, imagine a pigeon feeding in a field. As you saw earlier, ingestion of food will tend to increase drinking motivation. However, water is some distance away. At what point should it switch from feeding to drinking? Is it better, for example, to satiate the feeding tendency and then go to water? If not, how frequently could it afford to switch between feeding and drinking? There will be a cost attached to changing activity. There is the energy cost of flying as well as possibly an added risk of predation. Logically, one might suppose that the greater the distance between food and water, the less frequently should it switch between them. The bird should be prepared to accept the *cost* of displacement in fluid level because of the *benefits* that derive from not switching too often.

It is difficult to do experiments on animals in the wild but David McFarland (1985) and his associates in Oxford have been able to show evidence for such an effect even in the narrow confines of the laboratory. Food and water-deprived Barbary doves (*Streptopelia risoria*) were tested in the apparatus shown in Figure 7.17. Pecking at one key in the Skinner box delivers food and pecking at the other delivers water. To switch between the food tray and the water tray required the bird to negotiate its way around a barrier. As the barrier length was increased so the bird's tendency to persist with one activity (to 'lock-on') before switching to the other increased.

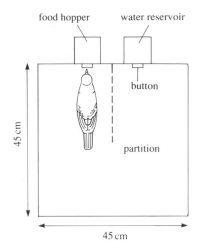

Figure 7.17 Apparatus used to study the timing of food and water intake by a Barbary dove.

Having discussed switching between activities and negative feedback, it is necessary to consider an issue that arises from the juxtaposition of the two topics. Suppose that an animal has been deprived of both food and water for a period of time and then allowed food and water. Suppose that the animal performs the behaviour appropriate to whichever motivation is the strongest, feeding or drinking. Assume that the strength of feeding motivation is just slightly larger than that of drinking motivation and so the animal commences feeding. By negative feedback principles, ingested food should lower feeding motivation and it would switch to drinking. A mouthful of water would then lower drinking tendency and it would switch back to feeding. Thus one might expect dithering. In fact what is observed under these conditions is a considerable persistence with one activity before a switch is made to another. How can this be explained in the language of motivational systems?

In a negative feedback system, action tends to decrease the underlying cause of the action. For example, dehydration motivates drinking but, because it decreases dehydration, drinking lowers motivation. However, there is evidence that in addition to negative feedback, a motivational system such as drinking incorporates positive feedback (see Figure 7.18). With positive feedback, rather than ingestion decreasing the strength of motivation, it actually *increases* it. The positive feedback effect is short-lived relative to the negative feedback effect. It gives the behaviour being performed a 'boost' which keeps the animal going until some other motivation can oust the first. In the longer term, negative feedback will dominate over positive feedback. So the long term effect, after about 5–20 minutes, will be one of negative feedback, i.e. from then on drinking lowers drinking motivation. The short-term effect, e.g. about 0–5 minutes, of drinking will be to increase drinking motivation. The effect will be that, once started on drinking, the animal will tend to persist with this. If it is simultaneously hungry, it will take longer before the command from the feeding motivation system is stronger than that from the drinking motivation system and feeding is able to take over.

Figure 7.18 Negative and positive feedback effects on drinking motivation

7.7 Conclusion

The central theme of this chapter is that behaviour is determined by many interacting factors. Consider, for example, the decision to feed or drink. This depends upon the relative strengths of the two motivations. As was discussed, drinking motivation depends upon the physiological state of the body fluids because of the negative feedback aspect of the system. In addition drinking can be influenced by feedforward effects, for example, in anticipation of the dehydrating effect of eating dry food. Evidence was given that feeding motivation is increased by the stimuli that in the past have been associated with food presentation. The motivational value of a given food can be increased or decreased by associations between the food and the consequences of ingestion. A decision whether to switch between feeding and drinking will depend not only upon the relative strengths of feeding and drinking motivation but also upon the cost associated with changing from one activity to the next.

The chapter has also emphasized the importance of both internal and external factors in determining any motivation. This was true of such diverse systems as feeding, sex and aggression. However, even after acknowledging this, we still have to exert caution in dichotomizing factors into external and internal. A particular food (external factor) might be of an intrinsically attractive quality but of low motivational value as a result of an association with illness (internal factor). Similarly for fear and aggressive behaviour, a previously neutral external stimulus might become aversive because of an association with trauma.

Although it is pointless to ask whether either external or internal factors are more important in determining motivation, *variations* in motivation might more often be associated with one or other factor. Consider a rat in a laboratory cage on a standard diet. The times at which it initiates and terminates meals might largely be explained by a fall and rise in some energy variable in its body. Contrast this with when a wild rat shows fear and aggressive behaviour. These times might be explained almost entirely by events in the external environment, for example, the appearance of a predator or a rival. However, these are two extreme ends of the spectrum. One can also find cases of where variation in aggressive behaviour needs to be explained also in part by considering variations over time of the internal factor of testosterone level, e.g. the amount of fighting among red deer (*Cervus elaphus*) stags at different times of year is known to be related to different levels of testosterone. Conversely, to explain variation in feeding, at times we need to appeal to external factors, for example, the arousal of feeding tendency in a 'satiated' rat by presenting a novel food or stimulus predictive of food.

It was argued that motivation is associated with both flexibility (e.g. various ways of getting to a given goal) and species-typical responses (e.g. the species-typical mating posture of female rats). At first sight, this might appear to be contradictory. In fact, there is no contradiction. Flexibility in being able to reach a given goal (e.g. to swim or to walk to food) is compatible with specific behaviours also being shown at a later stage. Thus a female rat might show flexibility in learning her way through a maze or pressing a lever in a Skinner box to obtain access to a male, but in the presence of the male might then show the stereotyped and species-typical mating posture (Figure 7.3).

To some extent this chapter contrasted psychological and ethological approaches. However, a complete understanding requires an integration of both. Traditionally, psychologists have been most concerned with studying individual systems, one at a time. For example, a researcher looking at feeding would wish to minimize the influence of any other motivation upon his or her data. To achieve this, typically a rat would be housed in a temperature-controlled environment and maintained with access to water *ad libitum*. As far as technically possible the rat would be set the single task of regulating its energy balance in response to challenges, but without 'contaminating' influences from other motivational systems. By contrast, ethologists have been more concerned with how decisions are reached when there is more than one competing task set the animal. The ethologist would acknowledge the insight to be gained by looking at individual systems but would emphasize that, in the natural environment, behaviour would rarely be the expression of a single motivational system. For instance, some of the set-points defined by the psychologist and presumably attained in the laboratory cage might never be reached in the natural environment, as was discussed for the lizard in the desert. There is every reason for more collaboration between psychologists and ethologists in the future.

Summary of Chapter 7

Motivation in relation to four aspects of behaviour were emphasized, as follows:

(a) Each motivational system has the potential to change its responsiveness to an external stimulus. Feeding and drinking do this mainly as a function of a physiological state critical to survival, sex to some extent as a function of hormones. Aggressive behaviour and defence show this property less than other systems, though aggressive behaviour can vary according to testosterone levels.

(b) Motivation was associated with the flexibility of behaviour, the ability of an animal to find a novel means to reach a goal.

(c) Behavioural decision making was discussed in terms of the animal making a choice between different activities.

(d) Motivation potentiates certain species-typical behaviour.

Motivations are determined by three major sets of factor: physiological variables (such as hormone level, energy state and fluid level), external stimuli (and stimuli associated with them), and learned associations between the stimuli and the consequences that follow from the animal's interaction with them.

Where there are clear fluctuations in motivation, it tends to be increased at times when it would tend to increase the animal's fitness. In some cases, the internal contribution to motivation is associated with a process of physiological regulation, e.g. feeding, drinking and temperature regulation. The motivation so aroused leads to behaviour that tends to restore the physiological state to its normal value. In so restoring it, motivation is decreased, an example of negative feedback. The roles of feedforward and positive feedback were also emphasized in the context of regulation. The function served by sex is not one of physiological regulation but reproduction. An example was given of where a fluctuation in the internal contribution to motivation corresponded to times when fertilization chances were maximum. Defence and aggressive behaviour serve to protect the physical integrity of the animal and to gain access to resources, such as territory or a mate.

Conditioning can reveal an important similarity between a variety of different motivational systems. In the case of feeding, defence and aggression, it was shown that a previously neutral stimulus acquires a motivational potency by being paired with a motivationally relevant stimulus. This was as true of the homeostatic system of feeding as of defence, sex and aggression. Conditioning also occurs between external objects and the consequences of interaction with these objects.

Objectives for Chapter 7

When you have read this chapter, you should be able to:

7.1   Define and use, or recognize definitions and applications of, each of the terms printed in **bold** in the text.

7.2   Describe how the concept of motivation is useful in the explanation of behaviour. (*Questions 7.1 and 7.2*)

7.3   Distinguish between changes attributable to motivation and to those attributable to: (a) learning, (b) maturation, and (c) injury. Distinguish between a motivational system and a simple reflex. (*Question 7.2*)

7.4   Describe for motivational systems, giving examples, some of the changes in responsiveness to external stimuli that are associated with internal changes. (*Questions 7.3 and 7.5*)

7.5   Describe and give examples of systems whose responsiveness to external stimuli changes little as a function of time, and yet fit other criteria for being classed as motivational. (*Question 7.4*)

7.6   Relate the function of motivational systems to the physiological processes underlying them and in so doing explain what is meant by homeostasis. (*Questions 7.5 and 7.6*)

7.7   Describe the main features of the body fluid system of an animal and relate these to the biological bases of drinking motivation. (*Question 7.7*)

7.8   Distinguish between negative feedback, positive feedback and feedforward. (*Question 7.8*)

7.9   Describe the relevance of classical conditioning to motivation, using feeding, aggression and defence as examples. (*Question 7.2*)

7.10   Explain what is meant by goal-directed behaviour and relate this to learning. (*Question 7.9*)

7.11   Give examples in which the maintenance of overall fitness requires that motivational systems sometimes fail to attain their optimal values. (*Question 7.10*)

Questions for Chapter 7

Question 7.1 (*Objective 7.2*)
Give an imaginary account of a sequence of sexual behaviour in the female rat, in which you illustrate the aspects of behaviour for which the term motivation is employed.

Question 7.2 (*Objectives 7.2, 7.3 and 7.9*)
A rat is maintained on food *ad libitum* and, on several occasions, placed in a maze containing food in one goal-box. It is observed to visit the goal box but not to eat any food. Later it is deprived of food and placed in the start-box of the maze. It immediately negotiates the maze to the food and eats. Give an account of these observations which includes the words *motivation* and *learning* so that their relevance to the observed behaviour is clear.

Question 7.3 (*Objective 7.4*)
For both observable behaviour and subjective sensations, describe how during a meal the reaction of a human towards food might change as a function of internal state.

Question 7.4 (*Objective 7.5*)
By the criteria of what constitutes a motivational system, how can defence and aggression be included in the category?

Question 7.5 (*Objectives 7.4 and 7.6*)
Could sex be described as a homeostatic system?

Question 7.6 (*Objective 7.6*)
In the context of (a) feeding, (b) drinking and (c) temperature regulation, explain what is meant by *negative feedback*.

Question 7.7 (*Objective 7.7*)
Suppose that the pathways that mediate negative feedback, from mouth and stomach, on drinking motivation could be cut or blocked. The rat is then injected with some concentrated salt solution. How would you expect the behaviour of the animal to differ from what it does when the pathways are intact?

Question 7.8 (*Objective 7.8*)
Give a brief account of drinking in which you demonstrate use of the terms feedforward and positive feedback.

Question 7.9 (*Objective 7.10*)
Imagine a modification of Dickinson's experiment (Section 7.5). Some water-deprived rats are trained to obtain water in a Skinner box. When the habit is well established the water is replaced by a salt solution. After a few sessions the rats stop pressing. They are then divided into two groups, A and B, both of which are maintained on water *ad libitum*. Group A is placed on a diet deficient in salt and group B is maintained on a balanced diet. Rats of both groups are then returned to the Skinner boxes under extinction conditions. In what way would the behaviour of group A and group B rats be expected to differ?

Question 7.10 (*Objective 7.11*)
Rats that are deprived of water reduce their intake of dry food even though their need for food does not decline. How might you account for this in terms of the fitness of the animal?

References

Cabanac, M. (1979) Sensory pleasure, *Quarterly Review of Biology*, **54**, pp. 1–29.

Dickinson, A. (1985) Actions and habits: the development of behavioural autonomy, *Philosophical Transactions of the Royal Society of London, Series B*, **308**, pp. 67–78.

Hollis, K. L. (1984) The biological function of Pavlovian conditioning: the best defense is a good offence, *Journal of Experimental Psychology: Animal Behaviour Processes*, **10**, pp. 413–425.

McFarland, D. J. (1985) *Animal Behaviour*, Pitman, p. 453.

Weingarten, H. P. (1984) Meal initiation controlled by learned cues: basic behavioural properties, *Appetite*, **5**, pp. 147–158.

Further reading

Colgan, P. (1989) *Animal Motivation*, Chapman and Hall.

Mook, D. G. (1996) *Motivation: The Organization of Action*, 2nd edn, W. W. Norton.

Toates, F. M. (1986) *Motivational Systems*, Cambridge University Press.

Toates, F. M. and Jensen, P. (1991) Ethological and psychological models of motivation—towards a synthesis, in J.–A. Meyer and S. Wilson (eds) *Simulation of Adaptive Behavior—From Animals to Animats*, MIT Press, pp. 194–205.

CHAPTER 8
COGNITION

8.1 Introduction

This chapter concerns the topic of *cognition*, and not surprisingly it is necessary to spend some time trying to define exactly what is meant by this term. Different authors differ widely in the way in which they use it. Problems of definition have occurred before, for example, in the learning and motivation chapters, but they are perhaps more intractable in the context of cognition. According to the dictionary, the word 'cognition' is defined as 'the action or faculty of knowing; knowledge, consciousness; acquaintance with a subject'. This chapter concerns the appropriateness of the expression 'cognition' to describe certain processes within the brains of non-human animals. Do they possess the faculty of cognition?

Beyond any controversy, it can be said that sensory information arrives at sensory organs, e.g. the eyes, ears and noses, of all animals, humans included. All animals exhibit behaviour. The question concerns the nature of the processes that lie between the arrival of sensory information and the execution of behaviour. Just how best can these processes be conceptualized? Are cognitive concepts useful? Cognition comes into the picture when one describes these processes in terms such as 'reasoning', 'showing intelligence', 'foresight' and 'synthesizing a novel solution to a problem from basic data'. Cognition is used when behaviour is more than simply a predictable and automatic consequence of stimuli that impinge upon the animal. It is a suitable term to describe the processes involved in making a choice as to the appropriate *strategy* for solving a problem. To varying degrees it can also involve the capacity of the animal to go beyond the raw sensory data it receives and to fill in the missing gaps. The term is used where information is held in memory over long periods of time and this information is then recalled and used flexibly in behaviour.

You have already met something of this topic in Chapter 7 where the purposive or goal-directed nature of motivational systems was discussed. To some authors, the term cognitive is appropriate to describe the information processing that is involved in generating flexible behaviour. That is to say it is information about the world that can be utilized in a variety of different ways by an animal in the course of its interaction with the world. A rat can use a rich variety of different means for getting to a given source of food. One might say that the knowledge 'food is located there' is held cognitively. This evidence is considered in more detail in this chapter. Though difficult to define, the meaning of the term cognition should become clearer with usage.

There are various shades of emphasis that can be applied to usage of the word cognition. Between incoming sensory information and behaviour there lie complex circuits of nerve cells with the capacity to generate seemingly intelligent behaviour and the term 'cognition' is sometimes used as no more than a means to capture the richness and complexity of this capacity. However, sometimes the term is also used to refer to the possibility that some animals have mental capacities similar to those of humans. The lay person often poses the question—do cats think? Are dogs conscious? This implies a particular personal, subjective and conscious quality to the cognitive processes but is inevitably difficult to define scientifically.

Try closing your eyes for a moment and let your thoughts wander. Focus upon the content of your thoughts. You are aware of a world of memories, private thoughts and feelings. You are the sole observer in this world and you are aware of yourself as a unique conscious being. Irrespective of your behaviour, this world of consciousness is meaningful to you. Do animals experience similar states of consciousness? Of course, humans do not know. However, they can speculate, and this chapter will briefly touch upon some of the kinds of experimental evidence that are appropriate to the discussion.

This topic is more than of just philosophical interest. It is of crucial importance to decisions concerning the relationship between humans and other animals. There can be no doubt that animals respond to noxious stimuli by attack and defence reactions. They respond to situations that humans would describe as fear-evoking by freezing or fleeing. It is certain that the brain structures underlying such behaviour are very similar to those parts of the human brain known to be closely associated with pain and fear reactions. The crucial question is whether animals have subjective conscious states of pain, fear, hope and depression, etc, that are in any way comparable to those of humans. Much of the current legislation on animal welfare in agriculture is based on the assumption that animals do suffer subjectively in a comparable way. Given that humans cannot ask them and expect an intelligent answer, the evidence that has been brought to bear on this question is of necessity indirect and some of it forms the content of the present chapter.

8.2   The experimental approach

8.2.1   Learning theory

Learning theory was introduced in Chapter 6 and discussed further in Chapter 7, where you met the famous experiments of Pavlov and Skinner, amongst others. These studies concerned such species as rats, cats, dogs and pigeons, though the implicit assumption was made that laboratory results can be extrapolated beyond these animals.

The experiment of Wagner as well as the evidence on biological constraints, both discussed in Chapter 6, are strong evidence in favour of the existence of complex information processing between sensory information and behaviour. These processes are often described by the term cognitive.

The most famous champion of the view that the behaviour of such animals as rats, cats and dogs can only be understood by using a cognitive concept was the American psychologist Edward Tolman who worked in the 1940s and 1950s. Tolman championed a particular view of the psychological processes underlying behaviour, with special reference to learning. Some of Tolman's most influential findings were obtained by looking at rats negotiating mazes. In such a situation, the animal's behaviour is *instrumental* in determining the outcome (Section

6.3.4). For instance, if it reaches the goal, it gets a pellet of food. In the case of rats learning their way through mazes, Tolman argued that learning consists of the storage of a particular kind of information in the brain—information concerning the relationship between objects and events in the external world. This is information of a kind that can then be utilized flexibly in the animal's purposive behaviour. Tolman argued that, in their performance of instrumental tasks, animals have a representation of the reward that awaits them in the goal box and this can be characterized as an expectation or, to use his term, *expectancy* of what is there. A number of bits of experimental evidence lead to such a conclusion. The reasons for Tolman's argument are best illustrated by the following examples of animal behaviour.

One of Tolman's students trained monkeys to perform an instrumental task to earn a piece of banana or lettuce. Both of these rewards are acceptable to monkeys but, not altogether surprisingly, banana is much preferred. Then an experiment was carried out in which a trick was played on the animal. The monkey was able to observe the experimenter place a piece of banana in the reward container. Then surreptitiously and out of sight of the monkey, the banana was replaced by a piece of lettuce. The monkey performed its instrumental task and was 'rewarded' with lettuce. It refused to eat the normally acceptable lettuce, but instead inspected the container. Occasionally, the animal would shriek in protest. The conclusion drawn from this study was that the animal's expectancy had been compared with reality and a disparity registered. This disparity triggered the negative reaction.

Subsequently a number of more formal experiments have lent support to this interpretation. Figure 8.1 shows an experiment in which two groups of rats were trained to find their way through a maze for the reward of food. One group was rewarded with mash and the other group with sunflower seeds. A general assumption in experiments of this kind is that, the higher the reward value, the faster the animal will go through the maze. Figure 8.1 shows that, given this assumption, rats preferred the mash to the seeds. On day 10 of the experiment, the rats that had until then received mash were switched to seeds.

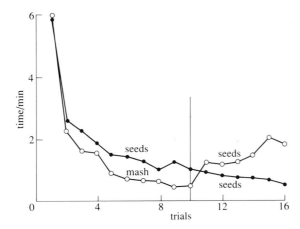

Figure 8.1   Result of an experiment in which rats were rewarded with either mash or seeds in the reward box of a maze. Note that time taken to run the maze is longer in the case of the rats rewarded with seeds, i.e. slower running speed. On day 10 and thereafter all rats were rewarded with seeds.

☐ Assuming that animals go faster through the maze for a reward of higher value, how did the value of seeds as reward compare between: (a) rats switched from mash to seeds, and (b) rats rewarded with seeds throughout?

■ The rats switched from mash to seeds went through the maze more slowly than those rewarded with seeds throughout, implying that the reward value of seeds was less for rats in condition (a).

Tolman's interpretation was that a failure to confirm the animal's expectation had a disruptive effect on its performance. Further experiments have pointed to the same conclusion and allowed some refinement of the argument, as described *below*.

In an experiment carried out by L. P. Crespi (1942), rats were divided into three groups, which were trained in a maze to receive the reward in the goal box of either 1, 16 or 256 small food pellets. This is the pre-shift period in Figure 8.2. For the two groups receiving the larger rewards, note the increase in running speed over the twenty trials. In the post-shift phase of the experiment, all rats received the same reward of 16 pellets.

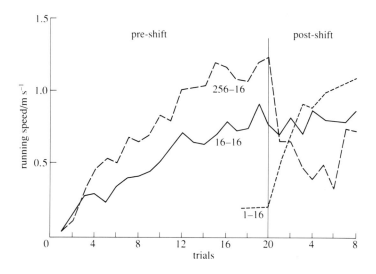

Figure 8.2 The results of Crespi's experiment. 1–16: rats receiving 1 pellet in the pre-shift period and 16 pellets in the post-shift period; 16–16: rats receiving 16 and 16 pellets, respectively, in the pre- and post-shift periods; 256–16: rats receiving 256 and 16 pellets, respectively, in the pre- and post-shift periods. Note that the figure combines data obtained from more than one experiment. The average running speed of 1–16 rats during the pre-shift period was approximately $0.2\,\mathrm{m\,s^{-1}}$, but detailed data on their performance in this period, like those presented for 16–16 and 256–16 rats, are not available.

☐ How would you characterize the behaviour of the rats in the post-shift period?

■ Rats shifted from 256 to 16 pellets ran slower than rats receiving 16 pellets throughout. Rats shifted from 1 to 16 pellets ran faster than rats receiving 16 pellets throughout.

This effect is termed the *Crespi effect* after the person doing the original experiment.

☐ How would you account for the Crespi effect in the cognitive terms developed so far in this chapter?

■ Rats developed an expectancy of the size of reward in the goal box against which the actual reward was compared. When the actual reward was worse than expected, this was reflected in a decrease in running speeds on subsequent trials. When the actual reward was better than expected, the running speeds increased.

Later theorists have referred to a 'frustration' effect when the reward is less than expected and an 'elation' effect when it is better. Further evidence on this subject can be derived from looking at the levels of certain hormones in the blood when an animal is exposed to such frustrating or elating circumstances. As will be discussed more fully in Book 5, when an animal is exposed to a situation that is described as stressful, a characteristic pattern of hormonal changes is seen. For example, the levels of a group of hormones called corticosteroids increase to greater than normal. Suppose an animal has been trained in a Skinner box and has learned to expect a certain size of food reward for pressing the lever. Then the reward size is changed. If the size of reward is decreased, the level of corticosteroids increases. If the reward size is larger than expected, a sudden decrease in corticosteroids to below the normal level is seen.

☐ What do these changes suggest about the organization of the nervous system processes that determine the rate of secretion of corticosteroids?

■ Detection of a disparity between actual and expected reward causes a change in the nervous control exerted on the secretion of corticosteroids. When the disparity is such that the actual reward is less than the expected reward, hormone secretion is elevated. When the opposite disparity is detected, i.e. the reward size is larger than expected, there is an inhibition of secretion.

Such results have led to the assumption that there is a similarity in the state induced by a reward being less than expected and the state of stress. Congruent with this is the observation that, when rats trained in Skinner boxes are placed under extinction conditions (Section 6.3.4), the goal box in which the reward was previously found, but in which it is no longer found, becomes aversive. Rats will behave as though trying to get out of the box. Thus it can be seen that cognition is intimately related to processes of motivation. When the consequences of an action are less desirable than expected there is a lowering of the motivation to continue the action. Such lowering is sometimes sufficient to cause the animal to suspend the action and to engage in some other activity. In other cases, as described above, the animal persists but with less vigour. In the case of extinction, one sees a steady decline in responding as behaviour fails to deliver positive consequences. Conversely, if the consequences of an action are better than expected then it is clear that there should be an increase in motivation, which will tend to cause the animal to persist with the activity in question.

Part of Tolman's argument was that an animal such as a rat forms an expectation of the quality and quantity of reward to be found in the goal box. In addition it forms a representation of the layout of the environment. As an example of this, take the maze shown in Figure 8.3a. The rat has to negotiate its way from A, through B, C, D, E and F, to arrive at the goal box G, where it is reinforced with food. Suppose that the apparatus is then changed to the 'sun burst' maze shown in Figure 8.3b. When the rat is released at A, it will find the way ahead blocked. It has a choice of arms 1 to 18. Typically, rats in such a situation tend to select that arm which points most directly to where the goal box used to be located, in this case either arm 5 or 6. In other words, the rats had learned something about the spatial layout of the apparatus and were able to *extrapolate* as to where the goal box would be, based upon the cues present in the sun-burst maze.

On the basis of such results, showing extrapolation about what is located where, Tolman proposed that, in the course of exploring an environment, an animal such as a rat constructs in its brain what he termed a **cognitive map** of that environment. This term can sometimes cause confusion. What Tolman meant was that information on the environment is stored in the brain in a form that has some features in common with a map. One should not think in literal terms of something like a street map of London being found in the head and the rat stopping to consult this: the term 'cognitive map' is merely a convenient term for the fact that information is held on the spatial relationships between parts of the environment. That is to say, a representation of the relative location of objects in the world is stored in the animal's brain. The rat can then utilize this information flexibly in moving around in the environment.

Another experiment of Tolman's gave further support to the notion of a cognitive map. The apparatus that he employed is shown in Figure 8.4. Three paths (A, B and C) lead from the start box to the goal box. By blocking various sections of the maze in turn, rats were forced to experience all paths during the initial stage of the experiment. When given a free choice, rats preferred to take path A, this being the shortest and most direct route. After a preference had emerged for path A, a block was placed at location X. What would be the rat's choice of route? The rat's preference now was for path B. The block was then imposed at Y rather than X.

☐ Which path would you now expect the rat to select?

■ You probably have become sufficiently convinced of the rat's cognitive capacities to answer correctly that it would tend to take path C.

☐ In what way might the term 'insight' be used to describe this result?

■ One might argue that the rat had extrapolated that Y, which blocked path A, would equally block path B but would not block path C.

The adaptive significance of the cognitive capacity is not difficult to appreciate. In their natural habitat, rats survive by successfully negotiating complex burrow systems and foraging for food in a variety of difficult and often hostile environments. The ability to form cognitive representations of the habitat that can be utilized in producing flexible behaviour involving the synthesis of information undoubtedly confers an advantage.

The meaning of the term cognitive map becomes still clearer when we consider some ingenious experiments carried out in the 1970s and 1980s by Richard Morris

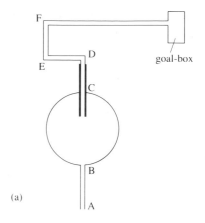

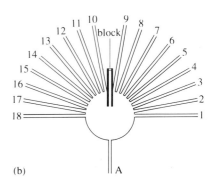

Figure 8.3 Maze employed by Tolman. (a) The maze in the learning phase, and (b) in the testing phase.

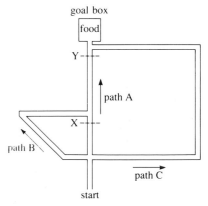

Figure 8.4 Maze employed by Tolman, which demonstrated insight in rats.

(1983), then at St Andrews University in Scotland. (Morris shares with Skinner and Tolman the rare distinction of having a piece of apparatus named after him.) As shown in Figure 8.5, the apparatus consists of a large bath of water that is made murky by the addition of milk. Submerged just below the fluid surface is a platform on which the rat can stand. Rats are good swimmers but will always climb out of water given the opportunity to do so. Therefore, on finding itself for the first time in the water, a rat will start to swim (Figure 8.6) in a manner that appears to be somewhat at random. In time, it will encounter the platform and climb onto it. The experimenter dries the rat in a towel and then puts it back into the tank, at some different point to where it was placed first time. This procedure is repeated a number of times, each time putting the rat in at a different location and observing the trajectory that it traces, as shown in Figure 8.7 (*overleaf*). Note that the rat's efficiency improves each time it is put back into the tank: it goes more directly to the platform. Morris argues that with experience, the rat builds up a cognitive map of the environment. Based upon landmarks outside the tank such as a bookcase, the light or the door, the rat can extrapolate to where the platform is located. It is not simply repeating a series of responses, since each time it is dropped at a new location in the tank and is required to find a new route to get to the platform.

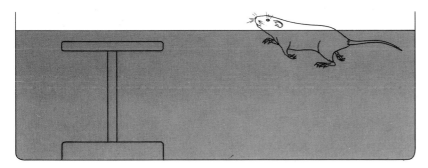

Figure 8.5   The Morris tank.

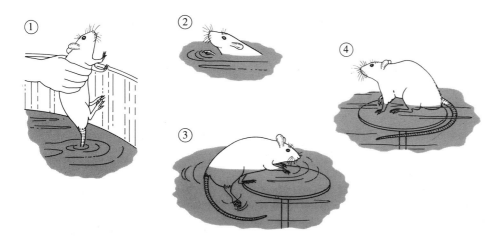

Figure 8.6   The Morris swimming task. (1) The experimenter places the rat in the tank, facing the outside wall. (2) The rat swims until (3) it finds the platform and (4) climbs onto it.

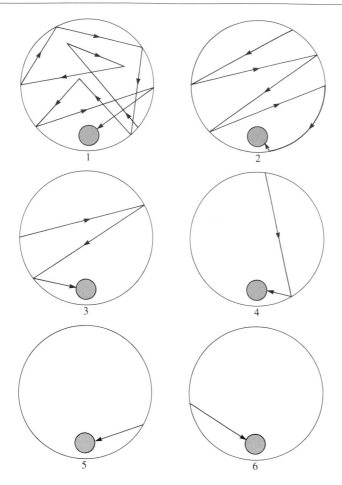

Figure 8.7 Routes traced out by a rat after being placed at various locations around the edge of the Morris tank. Note that on each of the subsequent trials (1–6) the rat's efficiency at swimming towards the platform increases.

When Morris removed the platform from the tank, the rat swam around as shown in Figure 8.8. It could be said that the rat is searching for the platform.

The evidence for cognitive maps in rats tested in laboratory conditions is paralleled by numerous findings on animals observed in more natural surroundings. In their foraging behaviour, a number of species can be shown to form representations of the type and quality of food to be found at certain locations within the context of a cognitive map of the environment. For example, a type of bird called the nutcracker (*Nucifraga caryocatactes*) is able to remember sites at which it has hidden food for months afterwards.

Sometimes when visiting a particular site, an animal will deplete the site of food. The best option is then to refrain from revisiting this site for a while, until it is likely that the food supply has been replenished. A type of Hawaiian honeycreeper, the amakihi (*Loxops virens*) will, following a visit to a flower and the depletion of its nectar, refrain from visiting it for several hours. This implies that part of the processes underlying the motivation to seek food consists of a memory of the time since the last visit.

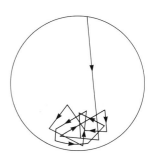

Figure 8.8 Route traced by a rat after removal of the platform.

☐   What is the functional significance of this cognitive capacity?

■   It would be a waste of time for the bird to revisit the flower before it had had time to refill with the nectar.

The conclusion is that mammals and birds store information in a way that can then be utilized flexibly in behaviour. In the next section, we look at further evidence that leads to a similar conclusion.

8.2.2   Declarative representations

As discussed above, both classical conditioning and spatial learning involving instrumental tasks lend support to the idea of animal cognition. Further evidence can be gained by looking at the behaviour of rats in Skinner boxes and asking how information is held in the animal's memory.

On considering how information is held in memory, modern cognitive approaches to learning draw a distinction between what are termed **procedural rules** and **declarative representations**. An example of a procedural rule is 'take the first turning right and then turn left at the first roundabout and keep going for 1 mile'. An example of a declarative representation is 'it is 100 m to the right of the central railway station'. The first case tells you what to do. The second tells you some information but not what to do on the basis of it. Procedural rules are rules of the kind that simply 'tell' an animal to act in a certain way in a specific situation. As an example, reconsider a phenomenon discussed in Chapter 7 and its implications for the goal-directed nature of motivation.

It was noted that, if an association is formed between a flavour and its consequences, this information can subsequently guide the animal's goal-directed behaviour in a different context. First, a rat was trained to earn pellets of food of a distinct flavour. Later it experienced nausea after ingesting food of this flavour in its home cage. Returned to the Skinner box and placed on extinction conditions, it pressed less than a control rat that had not had experience of pairing of taste and nausea. The argument was advanced that the animal's behaviour in the Skinner box was guided by a goal and the earlier experience of nausea had devalued the goal. This has implications for the form in which information is stored.

A procedural rule to encode nausea experienced some time after ingesting food of a particular flavour (X) would be of the kind 'do not eat flavour X'. Suppose that the animal has experienced nausea after eating this food in its home cage.

☐   What is predicted: (a) in the presence of food of flavour X in the home cage, and (b) under extinction conditions in the Skinner box, on the basis of storing information in the form of procedural rules?

■   In the presence of the food it would be predicted that the rat would not ingest it, since it would have learned the rule 'do not eat flavour X'. In the Skinner box, under extinction conditions, it would be expected that the rat would be unaffected by the experience outside the Skinner box. The information 'do not eat flavour X' should not influence its bar-pressing, i.e. the rule 'do not eat flavour X' does not say 'do not bar-press'. Since the bar-pressing behaviour of the rat *did* change, this suggests such information is *not* stored in the form of procedural rules.

☐   What is predicted on the basis of flexible goal-directed behaviour involving declarative representations?

■   It is predicted that: (a) the rat will refuse to ingest the food or will ingest relatively little, and (b) in the Skinner box, the rat might be expected either not to press at all or to press less energetically than an animal not having experienced an association between the characteristic flavour and nausea. Since this is what rats actually do, it suggests such information *is* stored in the form of declarative representations.

In practice, A. Dickinson (1985), who did this experiment, found that rats pressed relatively little when tested under these conditions. He argued that the rat stores two bits of declarative knowledge: (1) that the taste is followed by nausea, and (2) that pressing the lever in the Skinner box leads to the gain of food having this characteristic flavour. Note that neither of these bits of knowledge tells the rat what to do or what not to do. Rather, the rat must utilize this knowledge in its behaviour. One might express it that declarative knowledge confers the ability 'to put two and two together': lever pressing will yield unacceptable food.

In some cases, information is stored in the form of procedural rules in the central nervous system, i.e. as information on what to do. However, such information cannot be used flexibly. Consider, for example, an animal in the wild. Suppose that the information on an experience of nausea following ingestion were to be held in the form 'do not eat that food'. An animal would be able to refrain from ingesting the food in the future. However, it would not enable it to learn to refrain from approaching locations where such food is to be found, since the information only relates to one behaviour: ingestion. In fact, in such animals as rats, cats, dogs and primates, much, if not most, information is held in memory in a form that can be utilized flexibly and this involves declarative representations. A declarative representation encoding an experience would be something of the kind 'food of this flavour is bad'. Note that this piece of information does not tell an animal what to do or what not to do. It is simply a fact about the world. It needs to be integrated together with 'food of that flavour is over there' in order for the animal not to approach the source of this food.

8.3   Social intelligence in primates

The ability to 'put two-and-two together' in a variety of situations is perhaps seen at its richest and most advanced in the social life of primates. In 1976, an influential paper entitled 'The social function of intellect' was published by Nicholas Humphrey of Cambridge University, who summarized the essence of his argument as follows:

> '...the higher intellectual faculties of primates have evolved as an adaptation to the complexities of social living. For better or worse, styles of thinking which are primarily suited to social problem-solving colour the behaviour of man and other primates even towards the inanimate world.'

Humphrey's argument concerns the biological value of intelligence. For example, what advantage does it confer upon an ape to have the capacity for self-recognition, as revealed in the laboratory by the animal recognizing its own image in a mirror? Humphrey starts his analysis of the adaptive significance of intelligence by considering a central issue that we have discussed already (see Chapters 6 and 7)—that it confers an advantage to be able to predict what is likely to happen in the future. However, he notes a distinction between the intelligence required to perform two kinds of prediction:

> 'It requires, for instance, relatively low-level intelligence to infer that something is likely to happen merely because similar things have happened in comparable circumstances in the past; but it requires high-level intelligence to infer that something is likely to happen because it is entailed by a novel conjunction of events.'

The former kind of intelligence would seem either to correspond to, or have important features in common with, classical conditioning, discussed in Chapters 6 and 7. On the basis of a history of association between two events E_1 (e.g. a tone) and E_2 (e.g. a shock), on experiencing an event E_1, an animal acts in anticipation of E_2 happening.

This is the most simple kind of anticipation. Humphrey refers to the second kind of intelligence, being able to predict that something will happen on the basis of a synthesis of component bits of information, as 'creative intelligence'. He describes it as a characteristic of the higher primates and sees it as a process of 'invention'. By this term, he means the process whereby the animal can see new ways of using tools but can also discover novel behavioural strategies for reaching some desired state. As Humphrey admits, being able to manipulate the physical world—to avoid danger, find food, use tools to extract food from difficult locations—confers an adaptive advantage on an animal. However, he suggests that in practice, the role of invention has had a limited impact in this domain of primate evolution. Trial-and-error learning and imitation can account for much of this behaviour. Examples would include chimpanzees gaining food by unusual means, such as using a probe to pick up ants, and monkeys washing the sand off potatoes (as a group of Japanese macaques are known to do—see Chapter 10). Similarly, Humphrey notes:

> 'During 2 months I spent watching gorillas in the Virunga mountains I could not help being struck by the fact that of all the animals in the forest the gorillas seemed to lead much the simplest existence—food abundant and easy to harvest (provided they knew where to find it), few, if any, predators (provided they knew how to avoid them)—little to do, in fact (and little done), but eat, sleep, and play.'

So, Humphrey notes, biologists are faced with a conundrum. In the psychology laboratory, apes display a considerable capacity for creative reasoning and invention. Yet in the wild, in their interaction with their physical environment, the day-to-day tasks of existing seem to require nothing like such a capacity. Humphrey asks why higher primates need to be as intelligent as they are. Why have they intelligence apparently unmatched by most other species? He asks:

'What—if it exists—is the natural equivalent of the laboratory test of intelligence?'

In answer he writes:

'I propose that the chief role of creative intellect is to hold society together.'

To be precise, Humphrey emphasizes the vital role of what might be termed 'social invention'. This refers to the ability to predict and find novel ways of reacting to, and influencing the behaviour of, those conspecifics with which the animal interacts socially. How did Humphrey arrive at this conclusion? He spent some time in the early 1970s observing the rhesus monkey (*Macaca mulatta*) colony kept at the University of Cambridge research station at Madingley:

'I have looked anxiously through the wire mesh of the cages at Madingley, not only at my own monkeys, but at Robert Hinde's. Now, Hinde's monkeys are rather better off than mine. They live in social groups of eight or nine animals in relatively large cages, but these cages are almost empty of objects, there is nothing to manipulate and nothing to explore. Once a day the concrete floor is hosed down, food pellets are thrown in, and that is about it. So I looked, and seeing this barren environment, thought of the stultifying effect it must have on the monkey's intellect. Then one day I looked again and saw a half-weaned infant pestering its mother, two adolescents engaged in a mock battle, an old male grooming a female whilst another female tried to sidle up to him, and suddenly saw the scene with new eyes: forget about the absence of objects, these monkeys had each other to manipulate and explore. There could be no risk of their dying an intellectual death when the social environment provided such obvious opportunity for participating in a running dialectical debate. Compared to the solitary existence of my own monkeys, the set-up in Hinde's social groups came close to resembling a simian School of Athens.' (The meaning of the rather colourful expression 'dialectical debate' will be explained shortly.)

Humphrey argues that for any given member of a primate group there is a delicate balance to be struck between, on the one hand, preserving the stability of that social group and, on the other hand, exploiting and outwitting other members of the group to gain an advantage for itself. When one considers what is involved here, then the intellectual demand is considerable. An animal must be able to estimate the consequences of its own behaviour and to predict the behaviour of others. It must be able to calculate the relative benefits and costs of a given course of social interaction. The situation might be a rapidly changing one, with new information of relevance to the cost–benefit calculation arising by the second. Humphrey suggests that in such a context, social skill is virtually synonymous with intellect. As he expresses it so aptly:

'The game of social plot and counter-plot cannot be played merely on the basis of accumulated knowledge, any more than can a game of chess.'

As with chess, the player is confronted with novel situations that have never been experienced before. In such a situation, previous experience will undoubtedly be relevant but it is not enough. The ability to synthesize a novel solution is needed. The expression 'dialectical debate' refers to the fact that, in social interactions, both parties will have goals and anticipations and both will be trying to influence the behaviour of the other. Both will commonly need to change not only tactics but also goals as the social interaction progresses. The essence of such skill lies in being able to anticipate the future and revise the anticipation in the light of new experiences. Humphrey writes:

'It asks for a level of intelligence which is, I submit, unparalleled in any other sphere of living.'

A very similar line of argument to that just given was presented by Alison Jolly, who spent time in Madagascar studying socially-living lemurs. Lemurs are primates closely related to monkeys and now confined to Madagascar. Monkey social life is probably more complex than that of lemurs but there are some interesting similarities. As is also true of many monkey species, lemurs live in complex social groups, termed 'troops'. These groups are made up of both sexes and animals of various ages. Social bonds between individuals are built up and strengthened by various mutual activities, such as grooming, social play and care of infants. There is a great deal of social learning involved in functioning successfully as a member of such a troop. The characteristic behaviour to be expected of each other member of the troop needs to be learned. Jolly emphasizes the importance in such species of learning about social transactions. For example, in the laboratory, when confronted with a problem-solving task for the reward of food, various primates, ranging from lemurs to chimpanzees, try begging from the experimenter at first. In the wild, social influences on feeding have been observed. For example, macaques learn about new foods by observing the reactions and behaviour of members of their troop with whom they have a social bond.

A number of complex social interactions have been described and analysed in a book by Richard Byrne and Andrew Whiten (1988) of St Andrews University called 'Machiavellian Intelligence'. The logic underlying this title is as follows. In complex social interactions, a primate will sometimes attempt to influence the behaviour of another by means that in human society would undoubtedly be described as practicing deceit. Some of the classic studies that led to the notion of Machiavellian intelligence were made by Frans De Waal of the University of Utrecht, observing a colony of chimpanzees (*Pan troglodytes*) at Arnhem zoo. The richness of social interaction in such a colony can be illustrated by two examples of chimpanzee politics (to use the expression coined by De Waal) taken from De Waal's observations.

Two chimpanzee mothers, named Jimmie and Tepel, are sitting under an oak tree. Nearby, their two infants are playing and the oldest female of the colony, named Mama, is sleeping. Suddenly a fight breaks out between the two infants, involving hair-pulling and screaming. Jimmie attempts to control them but without success. Tepel then wakes Mama by poking her in the ribs. Mama gets up. Tepel then points to the two recalcitrant infants. Mama moves forward in a threatening manner, with her arms waving in the air and barking. The infants stop quarrelling. Mama then resumes her sleep.

In order to interpret fully this scenario, De Waal argues that one needs to know some things about the colony and the social interactions within it. First, Mama is the most 'senior' member of the colony, to whom all the others defer. Second, conflict between infants can sometimes result in conflict between their respective mothers. Thus in the situation described, one might see Jimmie and Tepel as being in something of a dilemma. This was solved by enlisting the help of a member of the colony who was beyond challenge. Such behaviour implies the cognitive capacity of being able to predict the likely consequences of various courses of action.

The second example that De Waal cites involves a chimpanzee named Yeroen who has hurt his hand as a result of a fight with another called Nikkie. The wound appears not to be serious but chimps walk on their hands and feet and Yeroen is observed to be limping. Closer scrutiny reveals that Yeroen only limps when he is in the field of vision of Nikkie. He walks normally when out of sight of Nikkie. This deception continues for almost a week.

☐ Can you suggest a possible explanation for this behaviour and what advantage it might confer upon Yeroen to behave in this way?

■ One might argue that the intention was to create the impression of serious injury. De Waal suggests that the advantage might be that Nikkie would be less rough in the future.

Possibly Yeroen had previously experienced more gentle treatment at the hands of Nikkie after being genuinely hurt. This example would seem to epitomize what would be termed deception in human social interaction: presenting false information to another in order to influence their behaviour in a way that favours the deceiver. A complex cognitive skill seems to be involved: the ability to predict the behaviour of another based upon the transmission of false information and the consequent cognitions of another.

A particularly amusing account of deception in a captive group of chimpanzees was described by Sue Savage-Rumbaugh and Kelly McDonald. One animal called Austin was bullied by another, called Sherman. However, Sherman had a particularly strong fear of the dark, not shared by Austin. Over the years, by means of deception, Austin had learned how to exploit Sherman's fear of the dark. To achieve this, Austin would slip outside and make noises on the wall of the sleeping room so as to simulate an intruder's attempts to break in. Austin would then surreptitiously slip back in and proceed to peer out as if trying to locate the source of the intrusive sounds. This induced fear in Sherman. So successful was this deception that it enabled Austin to stop Sherman bullying him. When frightened like this, Sherman stopped his bullying and would turn to Austin for comfort. In cognitive terms, Austin might be said to have the capacity to predict that his own behaviour would create the cognition 'intruder present' in the brain of Sherman.

It has been argued by Nicholas Humphrey that a primate living in a complex social environment must be able, in effect, to read the mind of another animal. This involves taking the intention of another animal into account.

Robert Seyfarth and Dorothy Cheney have suggested that 'monkeys make good ethologists'. The point is a serious one—there is something in common between the role of an ethologist and what a member of a monkey troop needs to do. The ethologist begins with a description of behaviour, recording observations on such things as grooming, vocalization and copulation, involving interactions between particular individuals. However, both ethologist and monkey also need to generalize to produce *abstractions*. In other words, they need to assimilate knowledge on the nature of relationships in a form that can be applied to individuals other than the ones on which the observations were made. The abstraction can then be used to extrapolate, to make predictions about novel individuals on first encounter. These predictions might prove to be wrong but even an inadequate prediction is doubtless better than no prediction at all. Part of Seyfarth and Cheney's argument as to why monkeys make good ethologists is:

'...because they appear to go beyond simple observation to make judgements about their fellow group members based on abstract features such as social relationships and the meaning of vocal signals.'

For example, monkeys are able to group vocalizations according to meaning and not just on the basis of the physical properties of the various sounds. In some situations, a vervet monkey (*Cercopithecus aethiops*) is able to determine that a given individual is not to be trusted in emitting a certain sound in relation to a given event. The monkey is then able to abstract the property of the individual's *unreliability* from the specific situation to the general. Thus, in other situations, in response to vocal signals of quite different physical properties, the individual is still not trusted. Another individual will be associated with the characteristic of reliability. To illustrate this principle, consider the fact that vervet monkeys have three different alarm calls according to the nature of the predator they have spotted—leopard, eagle or snake. The 'leopard alarm' call evokes the response of climbing into the trees, the 'eagle alarm' call causes them to look up whereas the 'snake alarm' call evokes a response of looking down. Suppose two individuals differ in the reliability with which they use these alarm calls, as illustrated in Figure 8.9 (*overleaf*). When individual A produces the eagle alarm there is always an eagle present but animal B occasionally uses it in the absence of an eagle just to gain an advantage. In future, when on first hearing the leopard alarm from A or B another member of the group might respond immediately to A's call but seek some kind of confirmation before responding to B's.

8.3.1   The concept of self

To what extent do individual primates have a concept of 'self' that is in any sense comparable to the concept possessed by humans? To what extent are such animals conscious? Until quite recently it was argued by many authors that the question is an impossible one to begin to answer. Even in principle there is no means of interviewing an animal. However, some interesting experiments, which provide a certain amount of indirect evidence on this point, were reported by G. G. Gallup (1977). They involved an animal's reaction to its image in a mirror. Sighted humans take the facility for self-examination that a mirror provides somewhat for granted. Interestingly, people who are born blind but later recover their sight following an operation at first treat their mirror image as being that of another person. Young children at first show 'other-directed' behaviour towards their mirror-image, treating it as a playmate. Self-recognition only appears at around the age of 2 years. Mentally-retarded people and schizophrenics sometimes show no capacity for recognition of a mirror-image as 'self'.

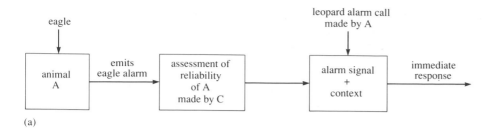

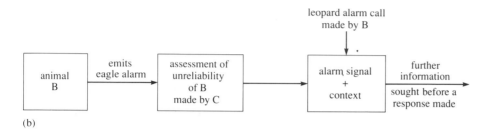

Figure 8.9 Difference in reaction to two monkey alarm calls. Phase 1—exposure to eagle alarm: in A's case in response to the presence of an eagle, and in B's case in the absence of an eagle. Phase 2—exposure to leopard alarm. Note that in phase 2, the animal is not responding simply to the physical property of the alarm but to this information in context, i.e. in combination with an assessment of the reliability of the transmitter.

Gallup placed a full-length mirror in the cages of individually-housed pre-adolescent chimpanzees, born in the wild. The chimpanzees' first reaction to the mirror was to behave as if the image were another animal and perform either threat or appeasement gestures towards it, as shown in Figure 8.10. By the third day, the animals' reactions had changed towards one of using the mirror to observe themselves doing various tasks such as rolling over, picking food from between the teeth, making faces or blowing bubbles. These are tasks that could not be observed otherwise (see Figure 8.11).

Then, while under the effect of a deep anaesthetic, red marks were placed on each animal, on its ear and eyebrow. The marks were made with an odour-free and non-irritating dye. The behaviour of the chimpanzees was observed for a period with the mirror removed in order to measure how frequently they spontaneously touched their ears and eyebrows. Then the mirrors were returned and the spontaneous frequency was compared with the animals' responses following return of the mirror. There was a dramatic increase in the frequency with which the chimpanzees touched their ears and eyebrows compared with when they were without the mirror, as Figure 8.12 shows. Chimpanzees were observed to examine and to smell the fingers that had been employed to touch the marks.

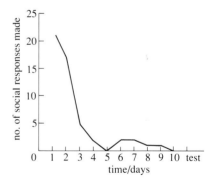

Figure 8.10 Number of social responses made by chimpanzees in the period of observation as a function of the day of test. 'Social responses' to a mirror image refers to 'other directed' behaviour such as threat or appeasement gestures.

☐ Why anaesthetize the animals before placing marks on them?

■ So that they couldn't know what had happened to them before the mirror experiment was carried out.

Another group of chimpanzees had red marks placed on them before ever being introduced to a mirror. These animals reacted like the first group, with hostility or appeasement towards their image. They did not show a high frequency of touching the red marks. It was argued that the first set of chimpanzees had formed the concept of self-image during their exposure to the mirror and that the red mark had changed this self-image.

Studies on other primates have also been carried out by Gallup. Olive baboons (*Papio cynocephalus*), gibbons, rhesus monkeys and stump-tailed macaques (*Macaca arctoides*) failed to show any evidence of self-recognition. After as long as 5 months of exposure to the mirror, a crab-eating macaque (*Macaca fascicularis*) showed no evidence of self-recognition. However, the animal was able to use the mirror as a tool to locate food that was otherwise out of sight. Chimpanzees and gorillas (*Gorilla gorilla*) are the only non-human primates known to show self-recognition.

Gallup speculated about the social function of self-recognition. He wondered whether rearing chimpanzees in social isolation would impair their capacity for self-recognition. Figure 8.13 shows the result of this experiment, comparing the performance of socially-reared and individually-reared animals in the red-mark experiment. The difference was dramatic. No self-directed behaviour was seen over the 10-day observation period in spite of the fact that the individually-reared animals showed intense interest in their image in the mirror throughout. The result suggests that development of the capacity for self-recognition depends upon social experience, but again caution is needed in drawing conclusions. Gallup explained the fundamental difference between monkeys and chimpanzees or gorillas in terms of the respective lack of and possession of a self-concept. He speculated that the former might be conscious but only the latter self-conscious.

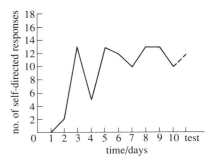

Figure 8.11   Number of self-directed responses made by chimpanzees in the period of observation as a function of the day of test. The expression 'self-directed responses' refers to such things as using the mirror to remove food from between the teeth or to touch parts of the body otherwise out of sight.

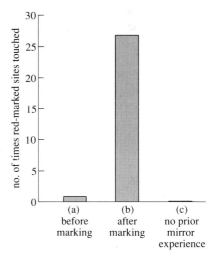

Figure 8.12   Number of times red-marked sites were touched: (a) before marking, and (b) after marking, by chimpanzees that had had previous access to a mirror. (c) Number of times red marks were touched by a group of chimpanzees who had never experienced a mirror before and who were then marked and given access to a mirror.

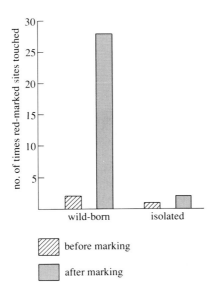

Figure 8.13   Number of times red-marked sites were touched before and after marking for wild-born, and therefore socially reared, chimpanzees and chimpanzees born in captivity and raised in isolation from other animals.

8.4   Conclusion

This chapter raises some of the most profound and challenging of issues in the behavioural and brain sciences. It started by considering the cognitive capacities of rats, as displayed in typical laboratory learning tasks. It was shown that even in such a highly restricted environment one can only explain what is happening by appealing to cognitive concepts such as expectation, representation and cognitive maps. The degree of sophistication of the brain processes becomes even more evident when the predictive capacities of primates are discussed. In their case, prediction can involve abstraction based upon general properties. Consideration of self-recognition in the final section raises several questions. For example, at what level of brain complexity does consciousness appear and are humans unique in their capacity for self-consciousness, i.e. to be conscious of being conscious? As yet these are questions that cannot be answered.

Summary of Chapter 8

The term 'cognitive' is not easy to define but is used to refer to the characteristics of certain types of brain processes that lie between incoming sensory information and the execution of behaviour. These are the processes that underlie such capacities as being able to show anticipation of the future, insight, and ability to extrapolate from data, amongst others. In a few special cases, the term is associated with consciousness. The evidence for the existence of cognitive processes is derived from various sources.

Edward Tolman and his associates performed a number of experiments that suggested the existence of cognitive processes. Animals showed signs of frustration when their reward was less than expected, suggesting a process of comparison between actual and expected states of the world. When negotiating their way in a maze, rats were able to extrapolate locations on the basis of limited information. Tolman introduced the term 'cognitive map' to describe the process underlying such an ability. Observation of the foraging strategies of some animals also suggests that representations of food availability are labelled within a spatial context.

Evidence from experiments using Skinner boxes demonstrates that rats can store information about the world in the form of declarative representations. This can then be used flexibly in the animal's interaction with the external world.

Cognitive capacities are particularly evident in the social skills of primate species. In their case, animals display behaviours that would be termed deceitful and Machiavellian in human societies. Certain primates show evidence of a self-concept which leads to the idea that the quality of conscious awareness may not be an exclusively human one.

Objectives for Chapter 8

When you have read this chapter you should be able to:

8.1   Define and use, or recognize definitions and applications of, each of the terms printed in **bold** in the text.

8.2   Give examples of some of the different ways that the term 'cognitive' is employed in the behavioural sciences, so as to illustrate its meaning. (*Question 8.1*)

8.3   Explain the broader significance of the term 'cognitive' to the way that animals are treated.

8.4   Describe the experimental evidence that supports the notion that, in instrumental tasks, animals form expectancies of the nature of reward. (*Question 8.2*)

8.5   Describe the motivational significance of an animal forming an expectancy of the outcome of instrumental behaviour. (*Questions 8.2 and 8.6*)

8.6   Describe what is meant by the expression 'cognitive map' and the experimental evidence that leads psychologists to postulate its existence. (*Questions 8.3 and 8.4*)

8.7   Describe the adaptive advantage to an animal in its natural environment of being able to synthesize information. (*Question 8.4*)

8.8   Describe the adaptive significance in terms of foraging of being able to form cognitive maps and label food sites in terms of such a map. (*Question 8.4*)

8.9   Describe the Morris swimming task and its implications. (*Question 8.5*)

8.10 Distinguish between, and give examples of, declarative representations and procedural rules. (*Question 8.6*)

8.11 Describe the social value of high-level primate intelligence.

8.12 Give examples of primate behaviour in which deception and the manipulation of other individuals' behaviour seem to be involved. (*Question 8.7*)

8.13 In the context of primate social behaviour, explain the meaning of the term 'abstraction'. (*Question 8.7*)

8.14 In terms of an experimental procedure designed to demonstrate it, explain what is meant by the term 'self-recognition'. (*Question 8.8*)

Questions for Chapter 8

Question 8.1 (*Objective 8.2*)
Explain why the term 'cognitive' is inappropriate to describe the processes underlying two simple reflexes, the pupil light reflex and the knee-jerk reflex.

Question 8.2 (*Objectives 8.4 and 8.5*)
Describe two measures by which imposing extinction conditions on an animal has effects similar to giving a reward that is inferior to what is expected.

Question 8.3 (*Objective 8.6*)
Hungry rats are trained in the maze shown in Figure 8.14a. From the start box A, they are given a choice of turning left or right. To the left they find a food reward and to the right the goal box is left empty. Once the rats are able to go straight to the food every time they are put in the maze, the experimental situation is changed, as shown in Figure 8.14b. The rat is now released from start box B. There are a number of distinct visual cues in the room. Which direction would the rat be expected to turn (towards goal box 1 or goal box 2) if: (a) it has a cognitive map of the maze and its surroundings, (b) it uses a procedural rule to find its way through the maze.

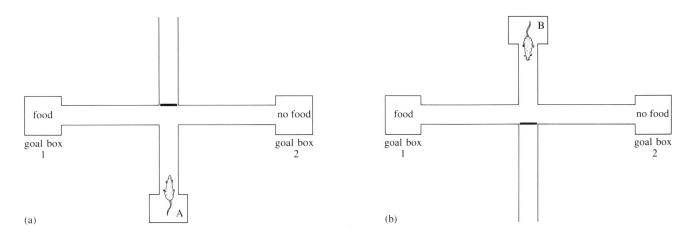

Figure 8.14   Maze: (a) training phase, and (b) testing phase.

Question 8.4 (*Objectives 8.6, 8.7 and 8.8*)
A tunnel used by a wild rat leads straight from its home territory to a food site, as shown in Figure 8.15. The rat is observed to use this route on most occasions when going to feed. However, it occasionally takes one of the two longer routes B and C. Describe what the adaptive significance of this behaviour might be. Describe a situation when the rat might be expected to use route B in particular or route C in particular.

Question 8.5 (*Objective 8.9*)
If Morris had not added milk to the water in his tank (Section 8.2.1), in what way would his argument have been weakened?

Question 8.6 (*Objectives 8.5 and 8.10*)
A healthy rat learns to earn two distinctively flavoured and equally attractive food pellets (A and B) in two Skinner boxes of very different physical appearance (X and Y). Neither of the food pellets contains any vitamins but they each have some nutritional value. Some weeks later the rats are made deficient in vitamin C. Then, in their home cages, they are given food having the characteristic flavour A but with vitamin C added. This corrects the vitamin deficiency. Some weeks later they are again put on a diet deficient in vitamin C. Given this information, how might you investigate whether information on diet and the instrumental behaviour they had learned to perform is held in a declarative or procedural form?

Question 8.7 (*Objectives 8.12 and 8.13*)
Considering a troop of monkeys, what would be meant by saying that member A had formed the *abstraction* that member B displays reliable information as far as vocal danger signals are concerned? How might such an abstraction be seen in the behaviour of animal A?

Question 8.8 (*Objective 8.14*)
Use of the group with no previous experience of a mirror in Gallup's experiment (group (c) in Figure 8.12) enables one to rule out *which* possible explanation for the effect shown by the experimental group?

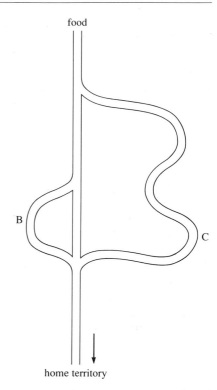

Figure 8.15   Tunnel system employed by wild rats.

References

Byrne, R. W. and Whiten, A. (eds) (1988) *Machiavellian Intelligence*, Clarendon Press, Oxford. (Includes papers by N. Humphrey, A. Jolly, S. Savage-Rumbaugh and K. McDonald, R. Seyfarth and D. Cheney, and F. De Waal.)

Crespi, L. P. (1942) Quantitative variation of incentive and performance in the white rat. *American Journal of Psychology*, **55**, pp. 467–517.

Dickinson, A. (1985) Actions and habits: the development of behavioural autonomy. *Philosophical Transactions of the Royal Society of London, Series B*, **308**, pp. 67–78.

Gallup, G. G. (1977) Self-recognition in primates. *American Psychologist*, **32**, pp. 329–338.

Humphrey, N. K. (1976) The social function of intelligence, in P. P. G. Bateson and R. A. Hinde (eds) *Growing Points in Ethology*, Cambridge University Press, pp. 303–317.

Morris, R. G. M. (1983) Neural subsystems of exploration in rats, in J. Archer and L. I. A. Birke (eds) *Exploration in Animals and Humans*, Van Nostrand Rhinehold (UK), pp. 117–146.

CHAPTER 9
REPRODUCTIVE BEHAVIOUR

9.1 Introduction

'Laymen get the impression that biologists have an inordinate preoccupation with sex. We are immoderate, and it is excusable: sexual behaviour is the key to understanding biological species.' (J. E. Lloyd)

The sexual behaviour of animals has been the subject of an enormous amount of research by ethologists and other kinds of behavioural scientists. This interest in sex stems, not from prurience, but from the fundamental importance of sex in the lives and evolution of most, but not all animals. As discussed later, some animals reproduce by other, non-sexual processes.

As emphasized in Chapter 2, sexual behaviour typically shows a very high degree of species-specificity; one sex performs displays, produces sounds, or secretes odours that are characteristic of their particular species, and the other sex recognizes and responds selectively to them. This specificity of stimulus–response relationships in sexual interactions maintains the integrity of species, and indeed it is an essential feature of how most species are defined. The specificity of sexual signalling systems has also made them attractive to biologists who seek to understand the relationship between external stimuli and behavioural responses, and the mechanisms in the nervous system that control behaviour. Sexual behaviour provides an excellent system for studying behavioural *mechanisms*.

Since the mid-1970s sexual behaviour has become the focus of increased attention by biologists interested in the *function* of behaviour. At that time, there was something of a revolution in the way that biologists viewed the evolution of social interactions between animals. Prior to this, biologists had commonly assumed, usually implicitly but occasionally explicitly, that the function of sexual behaviour was the perpetuation of the species. Thus sex was seen as an essentially cooperative exercise in which males and females interacted with one another towards a common goal, the production of offspring. Sexual behaviour, because it leads to the production of progeny, does have the *effect* of perpetuating species, but evolutionary biologists no longer see this as the correct way to understand the function of sexual behaviour, using function in its strict sense, as described in Chapter 4. There are two principal reasons for this change of view, one general, the other more specific. The general reason is that, as you will remember from Section 4.3, the theory of natural selection explains the characteristics of animals in terms of the benefits that those characteristics confer on *individuals*, not on *species*. The specific reason is that there are numerous examples where male and female animals behave towards one another in ways that are manifestly not cooperative. Certain female spiders and praying mantids eat males; in many species of birds males desert their mates and pair up with other females. The modern view of the sexual relationships of animals is that they contain elements both of cooperation and of conflicting interests. This theme will be developed later in the chapter, following a brief look at certain mechanisms involved in sexual behaviour.

9.2 Mechanisms underlying sexual behaviour

9.2.1 The timing of breeding

In the majority of animals reproduction is a seasonal activity, with mating and rearing of the young occurring at particular times of the year. At temperate latitudes, where there are regular and marked seasonal climatic changes, young are typically born in the spring. At tropical latitudes, breeding may occur at any time of year but often follows specific climatic events, such as periods of high rainfall. The timing of breeding activity and the way that it is linked to changes in the environment that affect reproduction can be understood in terms of various ultimate and proximate environmental factors. Ultimate factors are those features of the environment that determine when it is most adaptive for animals to breed, i.e. when breeding is most likely to be successful, in terms of the number of young reared. At temperate latitudes, most animals produce young in the spring and summer when food is most abundant and when environmental factors such as temperature impose the least stress on the young. Proximate factors are those features of the environment that provide the stimuli that elicit reproductive activity in animals. The most important of these are daylength and temperature in temperate habitats, rainfall and subsequent vegetation growth in arid, tropical habitats. The response of animals to proximate factors enables them to be in full reproductive condition by the time that ultimate factors become suitable for breeding.

☐ How does the distinction between ultimate and proximate factors relate to the different kinds of explanation of behaviour discussed in Chapter 1?

■ Ultimate factors relate to the consequences of breeding, i.e. the *function* of behaviour. Proximate factors are those that activate behaviour, i.e. its *causation*.

The importance of the timing of breeding to reproductive success can be seen by comparing the reproductive success of individuals that start to breed early or late in the year. Figure 9.1 shows the fledging success of great tits (*Parus major*) as a function of the time at which they lay their eggs.

For mammals, reproductive activity involves two major elements, mating and rearing the young. Whether or not these occur at the same time of year depends on the gestation period of the particular species concerned. The gestation period is the time between conception and birth; it is 9 months in humans, nearly 2 years in elephants, and 18 days in hamsters. In species with long gestation periods, mating and parental behaviour typically occur at different times of year and, as a result, they are controlled by environmental stimuli in rather different ways. This is well-illustrated by comparing two mammals, the red deer (*Cervus elaphus*) which mates in the autumn and calves in the early summer, and the wild rat, which mates and gives birth in the spring.

Sexual activity in red deer is triggered by the decreasing daylength that is characteristic of the autumn; they are described as a **short-day species**. A decrease in the duration of the light period stimulates a part of the brain, called the

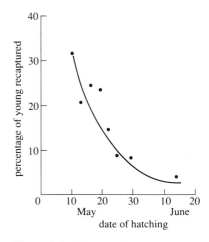

Figure 9.1 The post-fledging survival of young great tits in relation to their date of hatching. Survival is measured as the percentage of the young, hatched at each point in the season, that are subsequently recaptured alive.

pineal body, that in turn stimulates another part of the brain, the pituitary, to produce hormones that cause the gonads to develop and which activate sexual behaviour. In the wild rat, essentially the same hormonal and behavioural changes are triggered by the lengthening daylength characteristic of spring; they are a **long-day species**.

9.2.2   Hormones and sexual behaviour

Hormones are an extremely important influence on the behaviour of animals (see Chapters 2 and 5). Because hormones circulate around the body in the blood, they are naturally thought of as internal causal factors, but it is important to remember that many hormones, especially those involved in reproduction, are secreted in response to changes in the external environment, as described in the previous section and in Chapter 2.

You will remember from Chapter 2 that, in contrast to many external stimuli, which usually have a very specific effect on behaviour, hormones have more general effects, in two ways. First, they generally affect several aspects of behaviour, often relating to very different activities. Second, they tend to influence behaviour over a relatively long time-scale. Minute-by-minute changes in sexual behaviour, for example, cannot be related to simultaneous changes in hormone levels but must be explained as responses to more immediate events, such as the behaviour of the mate.

Hormones are internal stimuli that affect behaviour, and yet their action cannot be understood in isolation; it is essential to consider how they interact with causal factors in the external environment. As described earlier, hormone secretion is commonly a response to changes in the environment, such as daylength. Also, a hormone cannot usually activate behaviour in the absence of the external stimuli that elicit that behaviour (Chapter 2); animals obviously cannot show their full repertoire of sexual behaviour in the absence of a mate. One of the most complete analyses of the interplay between hormones, external stimuli and sexual behaviour is that carried out on the reproductive behaviour of doves, begun by Daniel S. Lehrman and carried on by many others (Silver, 1978; Cheng, 1979). This is summarized in Figure 9.2.

When a pair of doves are housed together with some nest material and a bowl in which to build a nest, they show a week-long period of courtship, towards the end of which they mate and build a nest. The female lays two eggs and male and female take turns to incubate them. By the time the eggs hatch, after about 2 weeks of incubation, both parents have developed the capacity to secrete a milky fluid in their crops, on which they feed their nestlings, which are called squabs. As shown in Figure 9.2, the behaviour of a female is influenced by several physiological factors, including four different hormones, and by external stimuli from the male, their nest, eggs and squabs. A feature of this system is that, at each stage, the female is stimulated to secrete the hormones that prepare her for the next stage. The male's courtship stimulates the secretion of oestrogen and progesterone; these not only cause her to become sexually receptive but are also necessary for nest-building and incubation. Incubation stimulates prolactin secretion in both sexes; this hormone is necessary, both for incubation to be maintained and for the secretion of crop milk. The secretion of prolactin both stimulates and is stimulated by incubation, an example of the process of positive

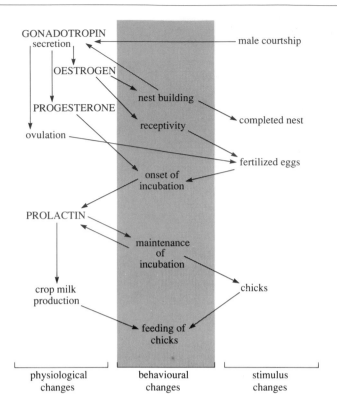

Figure 9.2 The principal interactions between physiological changes, external stimuli and behaviour involved in the reproductive behaviour of female Barbary doves (*Streptopelia risoria*). Hormones are shown in capitals.

feedback (see Chapter 7). This mechanism ensures that not only is incubation maintained until the eggs hatch, but also that, at that stage, both parents are secreting crop milk and are thus able to feed the squabs as soon as they hatch. The interaction between prolactin and incubation illustrates another important point. Hormones are generally thought of as influencing behaviour; this example shows that, in addition, behaviour influences hormone levels.

The control of the reproductive behaviour of the male dove shows similarities to, and differences from that of the female. Prolactin has a similar effect in both sexes. Testosterone shows a marked rise in the male after pair formation, but subsequent changes in the male's behaviour are responses to the female's behaviour, not to hormonal changes. He begins nest-building in response to the female spending long periods at the nest site, and not in response to either oestrogen, as in the female, or to a rise in testosterone.

9.3   Differences between the sexes

9.3.1   Some essential concepts and terminology

In everyday speech the word 'sex' encompasses many things, including the sex act (mating) and its preliminaries, conception and birth, relationships and differences between between males and females. For biologists, sex has a more limited, specific meaning and the other various aspects of reproductive biology have to be clearly defined and kept distinct.

'Sex' refers to a particular form of reproduction, sexual reproduction, that is differentiated from asexual reproduction. There are two essential features of sexual reproduction: the production (by meiotic cell division) of single cells, called gametes, that contain half the parental complement of genes, and the fusion of two gametes to form a zygote that will develop into a new individual (Chapter 3). Asexual reproduction typically involves neither gametes nor fusion; adults reproduce by dividing into two or by producing buds that become detached from the parent. In one common form of asexual reproduction, called parthenogenesis (meaning 'virgin birth') a single egg cell (produced mitotically or meiotically, depending on the species), divides and forms a zygote without fusing with a sperm. Some animals and plants are capable of both sexual and asexual reproduction, others reproduce by one or the other process. The evolutionary basis of the various forms of reproduction is not fully understood and the evolution of sex remains one of the major unanswered questions in biology. The most important difference between sexual and asexual reproduction lies in the difference in their genetic consequences. Animals and plants that reproduce by dividing or budding produce a number of progeny, called a clone, that are genetically identical to themselves. Parthenogenetic organisms produce progeny that are identical, or nearly so, depending on whether the eggs are produced by meiosis or not. Sexual reproduction produces progeny that are genetically very variable; they differ from both their parents and from one another (Chapter 3). The production of genetically varied progeny is widely regarded as a major function of sexual, as opposed to asexual reproduction and, therefore, a major reason why it evolved.

An almost universal feature of sexual reproduction is that there are two types of gamete, and progeny are produced by the fusion of two unlike gametes. There are a very few organisms that produce only one kind of gamete, two of which fuse to form a zygote—most produce gametes that exist in two forms, eggs (ova) and sperms, a condition called **anisogamy**. Eggs are relatively large as they contain the cellular material that provides the sustenance for the developing zygote, and they have little or no mobility. Sperm are relatively small, contribute only genetic material to the zygote and are very mobile. Anisogamy is an important phenomenon because it forms the basis for other differences between males and females, but the question of why and how it evolved is beyond the scope of this course.

In many species, including humans, individuals produce either eggs or sperm; individuals that produce eggs are, by definition, female, and those producing sperm are male. Along with the difference in the gametes they may produce, males and females in many animals differ in other ways; they may have different sex chromosomes, their behaviour is influenced by different hormones, and they show different patterns of behaviour. It is common practice to lump all these differences together under the heading of sex differences. These differences are sometimes regarded as aspects of *gender*, defined in the dictionary as the classification of animate beings and inanimate things as masculine, feminine or neuter. It is essential to be cautious in doing this, however. Consider, for example, animals such as earthworms and slugs that produce both eggs and sperm at the same time (they are called *hermaphrodites*), or those, like many species of fish, that can change from being male to being female, or vice versa, during the course of their lives. Here one might expect to find different hormones associated with male and female function, but individuals will not differ in their chromosomes.

Thus, while one aspect of 'gender' changes, another does not. The term 'gender' in this instance should strictly be applied only to male and female functions; it cannot be applied to the whole individual. As discussed later, this limitation in the use of the term is especially true of behaviour, because there is a great deal of variation in the way that various components of reproductive behaviour are divided between the members of a breeding pair.

Reproduction requires the expenditure of energy and the allocation of resources, such as food, to producing offspring; such energy and resources are, collectively, called **reproductive effort.** This effort is in addition to the energy and food that animals require to keep themselves alive and healthy, called **somatic effort.** For an individual animal, there is a certain balance between its somatic and its reproductive effort that maximizes the number of progeny it produces. Thus, an individual that puts a great deal of effort into reproduction may die of starvation or exhaustion, and one that puts a lot into survival may fail to reproduce. This balance between conflicting demands is called a trade-off. For example, red deer females breed for several years, giving birth to one calf at a time. If, in any given breeding season, they put too much effort into feeding their calf, they may be too thin to survive the following winter; the correct trade-off between reproductive and somatic effort in one year maximizes the chances of breeding in the following year. An important point here is that once energy and resources have been allocated to a particular activity, they cannot be recovered and used again for another.

Reproductive effort supports a number of different activities that together make up the process of sexual reproduction. The most important of these are: obtaining a mate, the mating act itself, and parental care of the young. Once again, these activities draw on a finite supply of energy and resources and there must be a trade-off between them; individuals that devote too much effort to obtaining mates may have insufficient reserves to rear their young. The reproductive effort that goes towards the care of the young is widely referred to as parental investment, a term that, by analogy with economics, incorporates the idea that parental effort yields some kind of return. The appropriateness of the analogy is strictly limited, however; whereas, in the financial world, money is invested to produce more money, in parental care energy and resources are expended to produce something quite different—surviving offspring. A more important limitation of the analogy with economics is that, unlike money, reproductive effort cannot be retrieved once it has been allocated to a particular activity.

An understanding of how individual animals allocate their effort among the various aspects of reproduction, making trade-offs between one and another, becomes important when trying to understand the different roles played by males and females in the reproductive process. Why, for example, is parental care carried out exclusively by the female in red deer, exclusively by the male in the stickleback, and is equally shared between the parents in the Barbary dove?

A final word on the terminology that is used in this area of biology. There is frequent use of words that are used in everyday language, a practice that can cause serious misconceptions. It was discussed *above*, for example, how the word 'investment' is used in a slightly different way in a behavioural context from its everyday use and it is necessary to be very specific about what is meant by 'sex' and 'gender'. The stickleback, because the male cares for the young, has been said to show 'sex-role reversal'. This is a pernicious use of words since it is based on the assumption that parental care is normally the prerogative of the female and that sticklebacks are therefore somehow unusual or abnormal. In fact, parental care by males is widespread among animals and it is seriously misleading to think of it as an unusual phenomenon. The term 'sex-role reversal' probably owes much to anthropocentrism, the practice of interpreting what animals do in terms of what humans do. (You no doubt have your own views on the validity of the assumption that, for humans, parental care is primarily a female activity!)

9.3.2   Fundamental and secondary differences between males and females

The fundamental difference between males and females is anisogamy; males produce tiny, usually mobile sperm, and females produce large, usually relatively immobile eggs. Due to the difference in the mobility of the two types of gamete, sperm typically have to move towards the stationary eggs, a factor that has a fundamental effect on the nature of the mating act. Even this generalization is not universally true, however; in seahorses, the male cares for the fertilized eggs in a special pouch on his body and the female delivers eggs to him through a penis-like organ (Figure 9.3). The definitive difference between eggs and sperms is their size.

Figure 9.3   A pair of seahorses mating; the female is on the left.

Eggs, because they contain a lot of cellular material (cytoplasm), are relatively large, and can survive for some time; because they are large, they are produced in relatively small numbers. Sperm, in contrast, are very small, short-lived (once they have become active) and are produced in enormous numbers. The number of progeny that a female can produce, her reproductive potential, is limited by the number of eggs she can produce; she cannot increase the number of her progeny by mating with several males. The reproductive potential of males, however, is virtually limitless and their reproductive success can be increased by mating with several females. On the basis of this difference we would expect, therefore, that females will typically devote most of their reproductive effort to the production, care and protection of eggs, but that males will devote most of theirs to finding and competing with other males for females. Some animals are like this—females mate with one male, males with several females—but very many are not, so clearly there are many more factors to be considered. The most important of these is parental care.

The eggs of many animals will not survive unless they are cared for and protected from predators. The eggs of birds, for example, must be kept at the correct temperature by incubation throughout their development and they must be guarded against a variety of egg predators, such as rodents, snakes, and other birds. For birds, reproductive success is limited by the need for incubation; an individual can only incubate as many eggs as it can fit beneath its body. Consequently, the reproductive success of many birds falls well below their reproductive potential; they can lay many more eggs than they can incubate. This fact is exploited in the poultry industry. If their eggs are taken away daily, hens go on laying more; if they are not taken away, a hen becomes broody, stops laying and starts to incubate as soon as she has built up a full clutch. (The ability of birds to lay more eggs than they can incubate has also proved important in conservation; several endangered species are being conserved by rearing their 'spare' eggs in incubators or using foster parents.)

For birds living in cold environments, where eggs will die if they are not incubated more or less continuously, or in environments where the risk from egg predators is high, breeding attempts may fail if a parent is not present at the nest all the time. Constant attention may also be essential after hatching, since the young often require frequent feeding, warmth and protection. For many bird species, one parent is insufficient to rear a brood of young successfully; when the parent leaves the nest to feed itself, or to gather food for the young, the eggs or chicks are likely to die of cold or through predation. In such circumstances, it is more adaptive for a male to remain with a female and share parental care with her than it is to allocate his reproductive effort to obtaining additional matings. No matter how many females he inseminates, if those females cannot rear their young alone, he will leave no offspring. Thus, hostile environments impose a selection pressure that favours males that remain with and participate in parental care with one female and selects against those that do not.

Even when the environment is not so hostile that a single parent cannot raise young on their own, it is to the advantage of each parent if the other shares in parental care, since the energetic demands of such care will be reduced. Assuming that, in birds, females are the primary care-givers to the young, what factors can enhance the probability that males will also care for the young? As explained above, the nature of the environment can be a potent external force in this respect, but there are also aspects of social behaviour that can play a part. Consider just two examples among many. First, it is quite common among birds that breed in dense colonies, such as gulls and many other seabirds, for females to show reproductive synchrony, such that they all mate and start to lay their eggs around the same day. (Chicks hatched at the same time as a lot of other chicks are less vulnerable to predation—see Chapter 10.) A consequence of this is that a male, having mated with one female, has a low probability of finding another female that has not mated. Thus, in the trade-off between staying with one female and seeking other matings, the benefits of seeking additional matings may not outweigh the costs, in terms of reduced hatching success, of leaving the first female. Secondly, if a male leaves a female, she may mate with another male, with the result that the male that leaves may father few if any of her eggs. When more than one male inseminates a female, there is competition among their sperm to fertilize the female's eggs. This sperm competition is an important factor in the evolution of sexual relationships. In some species, the sperm from two or more males simply mix in the female's reproductive tract so that, if two males inseminate a female, each fathers, on average, only half her progeny. In many species, however, the structure of the female's tract is such that sperm from the last male to inseminate her will fertilize all or most of her eggs. This phenomenon of 'last-male paternity', often makes it more adaptive for males to remain with a female that they have mated with and to guard her against the sexual attentions of other males than to seek matings with additional females.

These arguments incorporate only a few of the factors that have shaped the evolution of the many kinds of relationship that exist between male and female animals; limitations of space do not allow consideration of the many other factors that are involved. They do, however, make an important general point; the relationship between males and females involves both cooperation and a conflict of interests. Neither sex will leave any progeny if they do not cooperate to the extent of mating, at the very least. A very hostile environment may impose conditions under which young can only survive if both parents care for them. In other situations, however, one sex may behave in certain ways that alter the balance of interests of the other. Thus, breeding synchrony and the form of the female's reproductive tract may make it more adaptive for males to stay with a female than to seek further matings. When humans observe the attentiveness with which a male blackbird, for example, stays close to his mate, they tend to interpret this, anthropomorphically, as devotion to her. Instead, his behaviour is probably an adaptation by which he protects his paternity against the risk of her mating with another male. The interplay between cooperation and conflict of interests is discussed further in Section 9.3.4, with a specific example.

9.3.3   Mating systems

Across different species of animals, there is considerable diversity in the distribution of matings among individuals; the various kinds of distribution are called mating systems. The main categories of mating system are **monogamy**, in which individuals mate exclusively with one partner, over at least a single breeding cycle or season, and **polygamy**, in which individuals mate with several partners. Polygamy includes **polygyny**, in which an individual male mates with several females, but females mate with only one male, and **polyandry**, in which an individual female mates with several males, but males mate with only one female.

Before using these terms, some warnings are in order. First, the term 'system' can be misleading, since it implies some external constraint on the way animals behave. In human societies, monogamy may be prescribed or polygamy allowed or prohibited by legal and religious constraints, but among other animals the pattern of mating is a consequence of the behaviour of individuals. Second, while it is common practice to describe a *species* as monogamous or polygamous, these terms should properly be applied to *individuals*. Most sexual species have a 1:1 sex ratio (i.e. equal numbers of males and females), so only monogamy or polygamy shown by both males and females can be the rule for all individuals in a species. In species showing polygyny or polyandry, some individuals will have several mates, some one, and some none at all. Third, it is necessary to bear in mind the time-scale under consideration. In a species in which individuals breed several times, a female may be monogamous each year, but with a different male; over the course of her lifetime she is, in effect, polyandrous. Bearing these warnings in mind, the rest of this section considers the factors that promote monogamy in one species, polygyny or polyandry in another.

Monogamy

Monogamy is more common among birds than in any other group of animals and it is estimated that about 90% of bird species are monogamous; it is very rare among other vertebrates. The prevalence of monogamy among birds is probably due, as discussed earlier, to the fact that most of them produce young which, both as eggs and chicks, require considerable parental care if they are to survive. Eggs require constant incubation and defence against predators; chicks require warmth, protection and, in many species, continual feeding. Only in habitats in which it is warm, food is abundant, and predators are few is one parent able to rear the young alone. If females are assisted by a male, they can realise a greater proportion of their reproductive potential. If males engage in parental care of the young they have fathered, they gain greater reproductive success than if they do not.

Some birds maintain their pair-bonds over several years. In the kittiwake (*Rissa tridactyla*), the breeding success of individuals and pairs has been carefully monitored for many years and it has been possible to show that, the longer a pair stays together, the higher is their reproductive success. This provides an additional adaptive advantage for individuals that maintain a monogamous relationship.

Resource-based polygyny

This type of mating system occurs where males control access by females to resources, such as food or nest sites, that are essential to the reproductive effort of females, and compete to do so in such a way that the resources are divided unequally among males in a population. As a result, males that successfully defend a large share of resources attract and mate with many females; those that hold a small share obtain few or no matings.

If the resource occurs in patches, it is much more likely that there will be a pattern in which some males hold territories that are rich in the resource, while others will hold resource-deficient territories. A particularly clear example of resource-based polygyny is provided by a bird named the orange-rumped honeyguide (*Indicator xanthonotus*), of Nepal. In this species, beeswax obtained by raiding the nests of giant honeybees is an essential component of the diet of both sexes. Bee nests are scarce and occur only on cliff faces and so are patchily distributed. Males do not show any parental care but focus all their reproductive effort around bee nests, where a small proportion of males successfully establish territories. Only females and non-breeding immature birds are allowed into these territories and males copulate with all mature females that enter their territories to eat wax. Females are unresponsive to the courtship displays of non-territorial males.

Female-defence polygyny

In some species males compete for access to females directly, not by competing for resources, and successful males gather a group of females together. The formation of such groups is facilitated if females show a predisposition to gather in groups for some reason other than mating. For example, female elephant seals (*Mirounga angustirostris*) leave the water once a year, hauling themselves out onto a beach to give birth to their pups. Because suitable beaches are few and far between, large numbers of females tend to gather in confined areas. Mating occurs immediately after a female has given birth, at the same site, which thus becomes the focus of intense competition among males to gather groups of females. In the elephant seal, fights among males are extremely violent and only the largest and most powerful males are successful in gathering females. This intense selection for fighting ability has favoured large size in males, which are three times heavier than females. In a single season, only a third of males mate at all, and less than 10% of males fertilize nearly 90% of the females. Fighting incurs high costs for males; some die immediately after a season in which they have succeeded in mating and few are successful for more than 3 years, whereas females may breed for up to 10 years in succession. Some males never mate at all during their lives.

A male's success in maintaining exclusive access to a group of females depends not only on his ability to compete with rival males, but also on his ability to control the movements of females. Red deer stags spend all their time during the rut (mating period) competing with rivals by roaring and fighting, and in chivvying their hinds like a sheepdog to keep them in a compact group, but females frequently wander away and join other groups. The rarity of this kind of polygyny among birds is probably partly due to the fact that they are highly mobile and that males cannot easily control the movements of females.

Leks

A quite different form of polygyny occurs when males do not defend or control either resources or females. Instead, males gather in dense clusters, called leks, where they defend very small territories (Chapter 4). These contain no resources of importance to females but simply provide sites on which males display to and mate with females that visit the lek. Males display vigorously to females and it is typical of lek species that males have very elaborate plumage; the peacock is the classic example. Lek species have been studied very intensively, largely because of the opportunities they provide to study female choice (Section 9.4).

Polyandry

Polyandry, in which individual females simultaneously have long-lasting mating relationships with several males, is a very rare mating system. For this reason, and because it contradicts the general proposition that males are expected to be the polygamous sex, it has been intensively studied. It is exemplified by a water bird, the American jacana or lily-trotter (*Jacana spinosa*). Jacanas live on lily-covered lakes, where their greatly elongated toes enable them to walk on floating vegetation. Each female defends a large territory within which there are several smaller male territories, each containing a floating nest that is attended and defended by a particular male. She moves around her territory, mating from time to time with all the males, and laying eggs in each nest. Each male defends and carries out all the incubation of the eggs in his nest. The female is thus emancipated from parental care and allocates all her reproductive effort to mating, egg-laying and defending her territory against other females. She is considerably larger (50–75% heavier) than the males, is stronger than them and will stop any fights that break out between them.

The critical factor in the evolution of polyandry is the nature of the breeding habitat. Suitable breeding sites are scarce and nests are subject to heavy predation. These conditions favour the ability in females to lay large numbers of eggs, both to exploit such suitable habitats as do exist and to replace eggs that are lost to predators. Selection has thus favoured two aspects of female reproductive biology that enhance their egg-laying capacity, large body-size and a shift in resource allocation away from parental care and towards egg production. Selection for large size is accentuated by the advantage that it gives females in competition for breeding sites with other females. It has led to females being larger than males and thus able to control them.

This mating system is clearly highly advantageous to the female since she is able to achieve much higher reproductive success than if she had to care for her young. What are the selective pressures acting on males that have led to their behaving as they do? As discussed under monogamy, an environment that is hostile to egg survival will generally favour parental care by males, and polyandry most probably evolved from a monogamous mating pattern. A high level of predation requires the ability to replace lost eggs; only females can do this, shifting the allocation of parental care towards males and favouring large size in females. When the female is large enough to control the behaviour of the male, his options become restricted. To seek further matings he would have to leave his nest, exposing it to predation and himself to attack by the female.

There remains the question of why males within a female's territory tolerate one another. This is related to the pattern of mating and egg-laying shown by the female and introduces a critical factor in the evolution of this and other mating systems—certainty of paternity. The female mates with all the males in her group before and during egg-laying. Each male thus guards and incubates a clutch of eggs, only some of which are likely to be fertilized by him; on the other hand, some eggs that he has fertilized are being cared for by other males in the group. Natural selection will favour males that allocate paternal care only to those progeny that they themselves have fathered and will select against those that care for progeny that do not carry their genes. In the jacana, however, males cannot distinguish between eggs that they have fathered from those of other males, because particular episodes of mating by the female are not linked to particular episodes of egg-laying. Selection thus favours their caring for a clutch of eggs, some of which are probably theirs, and not disrupting the breeding effort of other males that are caring for eggs, some of which may also be theirs.

9.3.4 The dunnock: a variable mating system

Despite its drab appearance, the common dunnock or hedge sparrow (*Prunella modularis*) has a far from drab sex life. Meticulous studies of a population of dunnocks in the Cambridge Botanical Gardens by N. B. Davies (1985) have shown that there is considerable variation in the pattern of mating relationships among males and females. Some males are polygynous and have a territory containing two females and two nests with a brood, to each of which they provide some parental care. Some males are monogamous and devote all their parental care to a single brood. Others are polyandrous; each of two males cares exclusively for a single brood but the female also cares for the broods attended by both males. Davies recorded the number of young successfully reared from a large number of nests, according to whether their parents were polygynous, monogamous or polyandrous (Table 9.1). He also recorded the frequency with which individual males and females copulated with one another (individuals were marked with leg rings), so that he could estimate the certainty of paternity of males in polyandrous groups.

☐　From the data in Table 9.1, which mating system gives females the greatest reproductive success?

■　Polyandry. In polyandrous systems, females produced, on average, 6.7 fledglings.

☐　Which mating system gives males the greatest reproductive success?

■　Polygyny. On average, each polygynous male produced 7.6 fledglings.

☐　For each sex, how advantageous is monogamy, compared to the polygamous alternative?

■　For both sexes, it yields an average of 5.0 fledglings. For females this is better than polygyny, but not as good as polyandry. For males, this is better than polyandry, but not as good as polygyny.

Table 9.1   Reproductive success of male and female dunnocks in different mating combinations within a single population.

Mating system	Number of adults caring for young in each nest	Reproductive success (number of young fledged per season)	
		per female	per male
Polygyny (1 male, 2 females)	full-time female + part-time male	3.8	7.6
Monogamy (1 male, 1 female)	full-time female + full-time male	5.0	5.0
Polyandry (2 males, 1 female)	part-time female + full-time male	6.7	a 4.0* b 2.7

*In polyandrous groups, the female mates more often with one male (*a*) than the other (*b*); the paternity of *a* and *b* was estimated according to their relative frequency of copulation.

This example shows very clearly that there is indeed a conflict of interests between the sexes and that, within a single population, there can be a variety of mating relationships in which the balance of interests can tip towards one sex or the other. In the dunnock, the critical determinant of what pattern any one individual will be involved in apparently depends on his or her competitive ability. If a female is stronger than the males, she achieves polyandry; a very strong male will achieve polygyny. Birds of intermediate competitive ability that are well matched will become monogamous.

9.3.5   Experimental tests of mating system theory

The arguments presented in Section 9.3.3 about the environmental conditions that favour one kind of mating system rather than another are based on observational evidence. Researchers gather information about mating patterns and relate these to the nature of the environment in which different species live. These hypotheses can, however, be tested by carrying out appropriate experiments.

Polygyny

Mating system theory suggests that polygyny occurs where males defend resources that are essential to the reproductive effort of females and where there is variation among males in the quantity of such resources that they hold. This hypothesis can be tested by enhancing the resources held by selected individual males; if the hypothesis is correct, such males should attract more females. This kind of experiment has been carried out on an American bird, the red-winged blackbird (*Agelaius phoeniceus*), by Paul Ewald and Sievert Rohwer (1982). They placed either a single feeder containing protein-rich food, or several feeders, in the territories of particular males and compared the number of females that they attracted with the number attracted by control males that did not have a feeder. They found that males with a single feeder did attract more females than control birds, but that males with several feeders did not. What happened in the latter case was that several additional males moved in so that what had originally been a single territory containing several feeders became a number of very small territories.

☐   Do these results support the hypothesis being tested?

■ The results from the single-feeder territories do; males with enhanced resources in their territories did attract more females.

☐ What conclusion do you reach from the results obtained in the multiple-feeder territories?

■ They suggest that males are attracted to and compete for areas where resources are rich and which will, therefore, be attractive to females.

Monogamy

Mating system theory argues that monogamy is favoured when the environment is such that both parents can rear young more successfully together than if males are polygamous. This hypothesis has been tested in a number of birds by removing the male from an established breeding pair of birds after mating has occurred.

☐ If the hypothesis is correct, what result would you predict, in terms of the reproductive success of a female whose mate has been removed?

■ It should show a decrease, compared with that of females whose mate has not been removed.

In essence, this kind of experiment poses the question: 'are males necessary to females?' When a male-removal experiment was carried out on the American seaside sparrow (*Ammodramus maritimus*), it was found that the number of young successfully reared by experimental females was 33% less than the number reared by females whose males were not removed. In an experiment on house wrens (*Troglodytes aedon*) males were removed on four separate occasions; nestling survival decreased by 63% on one occasion, but showed no reduction on the other three occasions. The time when a reduction was found coincided with a period when climatic conditions were especially severe.

☐ Do these results support the hypothesis under test?

■ Yes they do. In both species, the presence of a male increases a female's reproductive success. In the house wren study, however, the variation in the results suggests that this effect only operates when the environment is particularly hostile.

Similar studies have been carried out on several species, and there is a great deal of variation between species in the strength of this effect (Figure 9.4), ranging from the seaside sparrow, in which male-removal leads to a decrease of 33% in the female's reproductive success, to the western sandpiper, in which the figure is 100%.

In some male-removal studies, researchers have closely observed 'widowed' females to see how, in behavioural terms, they compensate for the loss of their mate. For example, males were removed from female white-throated sparrows (*Zonotrichia albicollis*) when their chicks were 6 days old and it was found that females increased the frequency with which they left the nest to collect food and that, in consequence, they spent less time brooding their chicks. In this experiment, male removal did not reduce fledgling success but the chicks in experimental nests weighed significantly less than those in control nests.

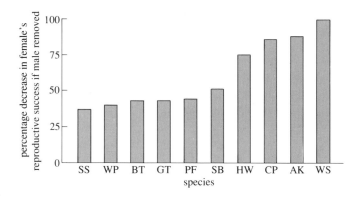

Figure 9.4 Maximum values, obtained from male-removal experiments, for the contribution of male birds to female reproductive success, for ten species. SS: seaside sparrow (*Ammodramus maritimus*); WP: willow ptarmigan (*Lagopus lagopus*); BT: blue tit (*Parus caeruleus*); GT: great tit (*Parus major*); PF: pied flycatcher (*Ficedula hypoleuca*); SB: snow bunting (*Plectrophenax nivalis*); HW: house wren (*Troglodytes aedon*); CP: common pigeon (*Columba livia*); AK: American kestrel (*Falco sparverius*); WS: Western sandpiper (*Calidris mauri*).

9.4 Mate choice

Since the mid-1970s there has been a great deal of interest in the question of whether animals choose their mates, that is: do they mate preferentially with particular individuals or are they equally likely to mate with any member of their own species? It is necessary to be careful here about what exactly is meant by 'choice'. It cannot be assumed that choice in animals involves the same kind of cognitive process as it does in humans when they choose a new car, for example, since, as humans, we are unable to observe directly the cognitive processes of animals, or even of ourselves (see Chapter 8). It is necessary to define 'choice' in terms of behaviour that can be observed. What is meant by 'mate choice' is any pattern of behaviour by an individual that makes it more likely to mate with a particular partner rather than another. There is a very important context in which almost all animals show choice in this sense: they mate selectively with members of their own species rather than with members of another species. Often, as discussed in Chapter 2, this is due to the sense organs of one sex being 'tuned' to the sexual signals produced by the other sex.

The hypothesis that animals do choose their mates is based on the following logic. First, males and females contribute in different ways to the reproductive process; most importantly, they differ in the extent to which they provide parental care and in the form of that care. Second, individual males and females will vary in their contribution to reproduction. Some females will be able to produce more eggs than others; some males will be better defenders of nests and providers of food than others. Third, the reproductive success of an individual will be greater if it mates with a partner that makes a large contribution to reproduction than it will if it mates with one that makes a small contribution. Fourth, natural selection will favour any mechanism in males and females that makes them more likely to mate with partners with a high reproductive potential.

It follows from this argument that the criterion on which mate choice will be based will vary from one species to another according to the exact nature of the contribution to reproduction that is made by males and females. Thus, in species in which males defend resources that are important to females, it is predicted that females will choose those males that defend the best resources.

☐ Can you recall two items of evidence, from earlier parts of this Chapter, that support this hypothesis?

■ (a) Female orange-rumped honeyguides mate only with males that defend bees' nests (Section 9.3.3), and (b) female red-winged blackbirds are more attracted to males holding territories in which the food supply has been enhanced by the inclusion of a food hopper (Section 9.3.5).

On the basis of the fundamental difference between males and females, anisogamy, it is widely argued that, because the reproductive potential of females is much less than that of males (Section 9.3.2), females should generally be more choosy than males. There are certainly many more examples in the literature of female choice than of male choice, but this may largely reflect a bias by investigators who took this assumption for granted. There is, in fact, increasing evidence that males do choose their mates. The remainder of this section looks, briefly, at evidence for both males and females that mate choice does occur, and the exact form that such choice takes.

9.4.1 Guppies

Guppies (*Poecilia reticulata*) are small fish, very popular with aquarists, in which the male is much smaller and more brightly coloured than the female; neither sex cares for the young. Males court females with a variety of postures in which they display their enlarged fins and bright colours. At first sight, they appear to be indiscriminate, but detailed analysis reveals that their tendency to display to females varies according both to the size of the female and to the phase that males have reached in their reproductive cycle (Figure 9.5). In addition, their various displays are performed at different frequencies. The most common (posturing) indicates a low level of sexual motivation; the least common (complete sigmoid) indicates high motivation. Study Figure 9.5 carefully.

☐ What effect does female size have on male courtship behaviour?

■ Whatever the male's internal state, larger females are more likely to elicit courtship displays than smaller females. Larger females are more likely to elicit high motivation displays (S).

This effect falls within the operational definition of mate choice given above. Because males are more likely to display to larger females, and because courtship is a prelude to mating, they are more likely to mate with larger females.

☐ Why do you suppose it might benefit males to mate preferentially with larger females?

■ Larger females carry more eggs, and so choosing larger females will increase a male's reproductive success.

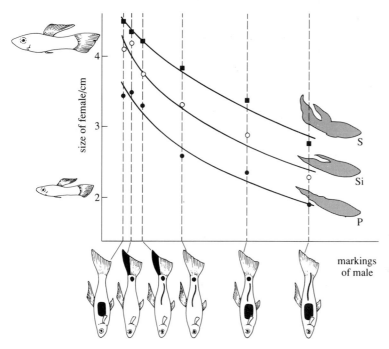

Figure 9.5 The influence of external stimulation (the size of the female) and male reproductive state (indicated by male colouration) on the courtship of male guppies. The markings of males change as they move from a state of low sexual motivation (on the left) to a state of high sexual motivation (on the right). Each point represents the minimum size of female necessary to elicit a particular display. Three displays are shown: posturing (P), sigmoid intention movements (Si), and sigmoid movements (S).

Male choice for larger, more fecund, females has been demonstrated in a number of species, especially among fishes, insects and amphibians, animals in which female fecundity is strongly correlated with body size.

Figure 9.5 illustrates another interesting effect.

☐   What effect does a male's internal state have on his behaviour?

■   Males whose state of sexual motivation is low respond only to the very largest females; those with high sexual motivation will respond to both small and large females, and are less discriminating.

This illustrates an important point made in Chapter 7, which discussed motivation. The tendency of a male guppy to perform courtship is the result of two factors, the strength of the external stimulus (female size) and the level of his internal motivation. A male with low motivation requires a strong external stimulus; one with high motivation requires only a weak stimulus.

Finally, this example illustrates the general point discussed in Chapter 1, that a given pattern of behaviour can be explained in different ways. Illustrated here is an interaction between internal state and external stimulus (a causal analysis), and the fitness benefits of choosing larger females (a functional analysis).

Female guppies also show mate choice; females are more likely to mate with males with several, large, bright red spots on their body than with males in which these spots are poorly developed. What adaptive basis this preference could have is not entirely clear but it is believed to be related to the fact that, to develop red colouration, fish require in their diet a particular class of nutrients called carotenoids. These are typically scarce in the environment and so only the most vigorous and healthy males can obtain them in quantities that are sufficient to make large red spots. Female preference may thus select for the most healthy and vigorous males in a population.

9.4.2 Sticklebacks

As in guppies, male sticklebacks court large, fat, more fecund females more vigorously than small females. For their part, mate-choice theory predicts that females should choose males according to their contribution to reproduction, i.e. parental care. The amount of time that male sticklebacks devote to defending and caring for their nests and eggs is proportional to the number of eggs that their nest contains; males with a lot of eggs are more attentive fathers.

☐ In the light of this observation, by what criterion would you predict that female sticklebacks should choose males?

■ They should choose males that already have a lot of eggs in their nest.

This prediction has been tested experimentally and has been supported; if eggs are added to a male's nest, he becomes more attractive to females. There is an interesting twist to this story; male sticklebacks have been observed to raid the nests of other males and to take eggs back to their own nest. As a result, they provide care for eggs that they have not fertilized themselves, but they may make themselves more attractive to females.

9.4.3 Female choice for male morphology

In many animals, breeding males are more brightly coloured than females and, in the breeding season, develop large, conspicuous display structures. Charles Darwin (1871) suggested that such characteristics, which are often apparently detrimental to male survival—they may make them more conspicuous to predators for example—evolved because females prefer more highly decorated males as mates. The peacock's train provides a classic example of a male character that is assumed to have evolved as a result of female choice. The peacock (*Pavo cristatus*) also typifies many species in which males are highly decorated, in that the male's sole contribution to reproduction is mating; the female cares for the young on her own. The hypothesis that females choose the most decorated males in a population has been tested in two ways, by observation and by experiment.

The observational approach involves measuring the development of male decorations in several individual males and scoring the mating success of those males. If Darwin's hypothesis is correct, the degree of development of a male's decoration should be related to his mating success. Figure 9.6 shows the results of such a study on peacocks, carried out by Marion Petrie and her colleagues at the Open University (Petrie *et al.*, 1991). The measure of male decoration in this case was the number of 'eye-spots' in a male's train, and the measure of mating success was the number of females mated with during a single season.

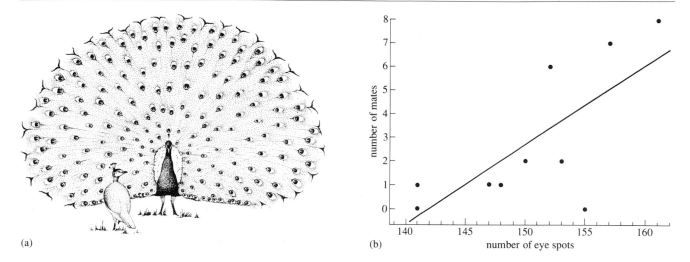

(a)

(b)

Figure 9.6   (a) A peacock displaying to a peahen. (b) The relationship between development of train (number of eye-spots) and mating success among a group of ten peacocks.

☐   Do the data presented in Figure 9.6 support the hypothesis?

■   Yes, they do; the more eye-spots a male has in his train (i.e. the more elaborate his train), the more females he is likely to mate with.

Correlational data of this kind provide support for the hypothesis but do not provide unequivocal proof. You will recall that, as pointed out in Chapter 2 of this book, correlation does not prove causation.

☐   Can you recall how, in the context of studying causation, hypotheses based on correlational evidence are tested?

■   By means of an experiment, designed to eliminate the possibility that a positive correlation between two variables is a consequence of their both being correlated with a third variable.

In the context of male decorations, it is necessary to eliminate the possibility that the greater mating success enjoyed by males with more eye-spots is due to their being superior in some other respect. This can only be done by means of an experiment in which the development of male decoration is manipulated. Such an experiment has yet to be carried out on the peacock but has been done with an African bird, the long-tailed widowbird (*Euplectes progne*) by Malte Andersson (1982). The male in this species possesses an enormously elongated tail (Figure 9.7). Andersson caught 36 males, having first recorded the number of nests in their territories (equivalent to the number of females). He then cut pieces of varying sizes off their tail feathers, and divided them into four groups:

(i)   Males whose tails were lengthened by replacing the cut portion with a longer piece cut from another male.

(ii)   Males whose tails were shortened by replacing the cut portion with a shorter piece.

Figure 9.7   A male long-tailed widowbird.

(iii)   Males whose tails were kept to their natural length by replacing the same piece that had been cut off.

(iv)   Males whose tails were left untouched.

The males were then released back into their natural habitat and closely observed, to record how many females each one attracted in addition to those already nesting in their territories.

☐   If the hypothesis that females prefer males with long tails is correct, what would you predict would be the relative mating success of the four groups of males?

■   Males in group (i) should attract the most females, groups (iii) and (iv) fewer, and group (ii) the least.

The results are shown in Figure 9.8.

☐   Do the results in Figure 9.8 support the hypothesis that long tails are attractive to females?

■   Yes, they do. Males in group (i), whose tails were lengthened, attracted more additional females than males in the other groups.

The results shown in Figure 9.8 are not as quite as clear-cut as they might appear. The differences between groups (ii), (iii) and (iv) are small enough to have arisen purely by chance.

☐   Why do you suppose that Andersson included group (iv) in his experiment?

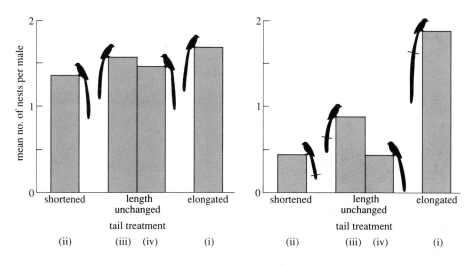

Figure 9.8   Results of Andersson's experiment in which the length of male widowbirds' tails were manipulated. *Left*: the mean number of nests in the territories of four groups of nine males before manipulation of their tails. *Right*: the mean number of additional nests in the territories of the same males after manipulation of their tails. The four kinds of tail manipulation, (i)–(iv), are explained in the text.

■   To eliminate the possibility that the actual procedure of having its tail cut and put back together might have affected a male's attractiveness, quite independently of whether it was made shorter or longer, i.e. it allows for the effects of the treatment.

In a similar experiment on a related species, the shaft-tailed widowbird (*Vidua regia*), Phoebe Barnard (1990) tested the behavioural responses of females to males with lengthened, shortened and normal-length tails in an aviary. Females showed a higher frequency of a behaviour pattern called solicitation to males with lengthened tails than to those with normal-length tails, and more to males with normal-length than to those with shortened tails. An intriguing result that emerged from this study is that males whose tails had been experimentally lengthened displayed and called more vigorously than those with normal tails.

The functional significance of female preference for decorated males in species in which males make no contribution to reproduction, other than mating, is unclear and controversial. One school of thought is that highly decorated males are in some way superior to less-decorated males and that, by choosing them as mates, females may endow their progeny with superior alleles. Another view is that female preferences are essentially arbitrary, and that males have evolved characters that exploit such preferences.

☐   Can you recall, from Chapter 2 of this book, a feature of sensory systems that might explain why females prefer very decorated males?

■   The supernormal stimulus effect. Just as the incubation responses of some birds are more strongly aroused by abnormally large eggs, so females may be more sexually aroused by males with especially elaborate plumage.

9.4.4   Choice for complementarity

As mentioned briefly in Section 9.3.3, monogamy is advantageous in several species of birds that breed many times during their lives, because the breeding success of pairs improves over successive breeding seasons. Long-term studies of the kittiwake, a cliff-nesting gull, have shown that this effect is more apparent in some pairs of birds than others. Some pairs, in a given year, fail to rear any young at all; this may be due to one partner being sterile or it may be that they do not coordinate parental care as successfully as other pairs. Among kittiwakes, it has been found that pairs of birds that are successful in one year are more likely to breed together in the following year than those that are unsuccessful. Unsuccessful birds often 'divorce' after an unsuccessful season and obtain new mates in the following season.

9.5   Parental behaviour

As discussed earlier in this chapter, courtship and mating are often only the beginning of the reproductive process. For many species, the greater part of the reproductive effort of males, of females, or of both is expended in parental care. Earlier sections also argued that parental care, and the way that it is divided between the parents, is a major determinant in the evolution of sexual relationships in animals. This section looks briefly at one aspect of parental care, the opportunities that it provides for learning by the progeny.

In the stickleback, parental care is provided solely by the male. He fans the eggs to keep them well aerated and defends them, and the fry into which they hatch, from predators. As they develop and grow, the fry swim out of the nest and explore their surroundings. Their father chases after them, sucks them up into his mouth and spits them back into the safety of the nest. As the fry get older, they become increasingly adept at escaping from their father. The effect of this early experience on their adult behaviour has been investigated in experiments in which young sticklebacks are brought up without a father. It was found that 'orphaned' sticklebacks, when adult, were much less wary of, and less able to escape from a predatory fish, such as a pike, than those that had received paternal care. Their experience of being chased and caught by their father influences the subsequent responses of sticklebacks towards predatory fish.

☐   What other examples can you recall from earlier in this book of adult behaviour that is influenced by early experience of parents?

■   You have read about two examples, the development of bird song and imprinting.

In many songbirds in which the father cares for the young, males learn the characteristics of their species' song from their father. As there are individual differences in the way that males sing, they also learn those aspects of the song that are particular to him and, in the process, they incorporate some particular features of their own. As a result, the song of an individual male is unique but contains features that resemble his father's song, as well as features that differentiate his species' song from that of other species. These variations can be important in adult life. For example, females may base their mating preferences on the nature of the songs of prospective mates, preferring males that do or do not sound like their father. Sexual imprinting in precocial birds such as ducks and poultry may also influence adult mating preferences.

Variation in the nature and quality of parental care, and its consequences in later life, has been especially well studied among primates. In some early and rather crude experiments, Harry Harlow (1963) reared young monkeys without a mother and found that this seriously affected their adult behaviour. They were unable to form normal relationships with other monkeys, they tended to be very fearful and aggressive, and they were very inept at rearing their own progeny. This effect, by which individuals that had been deprived of maternal care themselves became inadequate mothers, is called a transgenerational effect. It is an important developmental effect on behaviour which must be added to genetic factors and learning as a source of variation in the behaviour of animals.

The rather catastrophic results obtained by Harlow stimulated subsequent studies of parental care in monkeys, notably by Robert A. Hinde and his co-workers (Hinde and Spencer-Booth, 1971). These looked very closely at mother–infant relationships and found that there is a great deal of variation. Infant monkeys seek contact with their mothers, and mothers seek contact with their infants; as the infants get older the frequency of such contacts tends to decline. Mothers vary in the frequency with which they initiate contacts and in their response to infant-initiated contacts; some mothers reject their infants more often than others. By

236

looking at several mother–infant relationships, Hinde identified a continuum of mother–infant relationships with two extreme patterns, where mothers could be identified as 'rejecting' or 'possessive' (Figure 9.9). Most mothers lie between these extremes and are labelled 'controlling'.

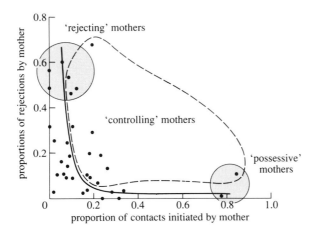

Figure 9.9 Qualities of mother–infant relationships between rhesus monkey (*Macaca mulatta*) mothers and their 16-week-old infants.

These variations in the quality of the mother–infant relationship have subtle effects on behaviour. For example, when infant rhesus monkeys are briefly separated from their mother, they show considerable distress. This distress is greater in the infants of rejecting mothers than in those of possessive or controlling mothers. According to Hinde's interpretation, infants with a more *secure* relationship with their mother before the separation will be less distressed after the separation.

9.6 Conclusion

In this chapter, you have looked, rather briefly, at various aspects of reproductive behaviour. This is a category of behaviour that is rich in diversity and it is possible here only to hint at the complexity of the subject. What you have read is intended, however, to make a general point about behaviour that is particularly apparent in the context of reproduction: animals show individuality. There is a tendency, at all levels of analysis, to make assumptions about the behaviour of categories of animals. It is often assumed, for example, that the much-studied rat is representative of mammals in terms of its behaviour and its physiology. It is not; a rat is a rat and an elephant is an elephant and we understand them better by considering their differences as well as their similarities. In the context of reproduction, you have seen that there are profound differences between the sexes. Male and female sticklebacks have much in common; they feed the same way and are attacked by the same predators but, when it comes to reproduction, they behave very differently. These differences arise from the fundamental difference between male and female gametes. It is important also, in recognizing fundamental and secondary sex differences, not to make assumptions across species. The stereotype of the aggressive, randy male and the nurturing, coy female simply does not apply to the great majority of animals. Biologists have to consider each species on its own, analysing its reproductive behaviour in terms of differences between sexes that arise from the fundamental difference between them but which are not prescribed by it.

Finally, you have seen that individuals within one sex vary in their reproductive behaviour. An understanding of individual differences is of profound importance in the study of behaviour, for two reasons. First, it is at the level of the individual that natural selection acts and biologists can therefore only fully understand how natural selection has shaped behaviour by studying the way behaviour varies at the individual level. Second, you have seen that, particularly in the contexts of mate choice and parental care, animals interact with each other as individuals; they do not react to all members of their species in the same way.

Summary of Chapter 9

Animals typically breed at particular times of year and their breeding activities tend to be timed such that reproductive success is maximized. This timing is brought about by their response to environmental cues, such as daylength, which causes the secretion of reproductive hormones. Hormones have a wide range of effects on the physiology and behaviour of animals and behaviour can have a reciprocal effect on hormone secretion (positive feedback). During behavioural interactions between males and females, the behaviour of each sex has a stimulatory effect on, and is stimulated by the behaviour of the opposite sex.

Sexual reproduction produces offspring that differ genetically from their parents and from one another. There are many differences between the sexes, in terms of morphology, chromosomes, hormones and behaviour, but the fundamental difference is anisogamy, the difference in the form and size of their gametes. Among species, there is a great deal of variation in the extent to which males and females show behavioural differences. In particular, species vary in the extent to which males, females, or both sexes care for the young. These differences are largely attributable to environmental factors.

The pattern of mating within a species is called the mating system of that species. The main categories of mating system are monogamy, polygyny and polyandry. The nature of the mating system in a particular species is determined largely by environmental factors, particularly the abundance and distribution of resources necessary for reproduction, but other factors such as the timing of mating and certainty of paternity are also important. Within a species, a diversity of mating systems may be found, as in the dunnock. Theories about the evolutionary factors promoting particular mating systems can be tested experimentally.

Animals of both sexes are expected to choose their mates on the basis of variation in the performance of potential partners in a number of reproductive activities, including their fecundity, their capacity to care for progeny, and their appearance.

During the time that young are cared for by their parents they acquire aspects of behaviour that are important later in life. Many birds learn the characteristics of their species' song. The adult behaviour of monkeys can be seriously impaired by disruption of their relationships with their parents.

Objectives for Chapter 9

After reading this chapter, you should be able to:

9.1   Define and use, or recognise definitions and applications of each of the terms printed in **bold** in the text. (*Question 9.1*)

9.2   Explain why sex is typically a seasonal phenomenon. (*Question 9.2*)

9.3   Describe the basic effects of sex hormones on behaviour.

9.4   Explain the evolutionary origins of differences between the sexes in terms of their behaviour. (*Question 9.3*)

9.5   Describe the fitness costs and benefits associated with different mating systems, and the importance of environmental factors in determining what mating system is adopted. (*Question 9.4*)

9.6   Explain mating systems as the product of the behaviour of individuals. (*Question 9.5*)

9.7   Recall observational and experimental evidence that animals choose their mates.

9.8   Give examples of how variation in parental behaviour can influence the adult behaviour of animals. (*Question 9.6*)

Questions for Chapter 9

Question 9.1 (*Objective 9.1*)
Give definitions of the following terms: (a) short-day species, (b) anisogamy, (c) polygyny.

Question 9.2 (*Objective 9.2*)
Red deer mate in the autumn in response to shortening daylength and give birth early the next summer when food is most abundant. Which part of this statement relates to: (a) ultimate factors, and (b) proximate factors that control their reproduction?

Question 9.3 (*Objective 9.4*)
Explain briefly why anisogamy leads to males and females showing differences in their reproductive behaviour.

Question 9.4 (*Objective 9.5*)
Name two aspects of food supply that can promote polygyny in a given species.

Question 9.5 (*Objective 9.6*)
Some male pied flycatchers have two mates and two nests at the same time. Is it correct to describe pied flycatchers as a polygynous species?

Question 9.6 (*Objective 9.8*)
What is meant by a transgenerational effect?

References

Andersson, M. (1982) Female choice selects for extreme tail length in a widowbird, *Nature*, **299**, pp. 818–820.

Barnard, P. (1990) Male tail length, sexual display intensity and female sexual response in a parasitic African finch, *Animal Behaviour*, **39**, pp. 652–656.

Cheng, M.–F. (1979) Progress and prospects in ring dove research: a personal view, *Advances in the Study of Behavior*, **9**, pp. 97–129.

Darwin, C. (1871) *The Descent of Man, and Selection in Relation to Sex*, John Murray.

Davies, N. B. (1985) Cooperation and conflict among dunnocks, *Prunella modularis*, in a variable mating system, *Animal Behaviour*, **33**, pp. 628–648.

Ewald, P. W. and Rohwer, S. (1982) Effects of supplemental feeding on timing of breeding, clutch-size and polygyny in red-winged blackbirds *Agelaius phoeniceus*, *Journal of Animal Ecology*, **51**, pp. 429–450.

Harlow, H. F. (1963) The maternal affectional system, in B. Foss (ed.) *Determinants of Infant Behavior, II*, Methuen, pp. 3–33.

Hinde, R. A. and Spencer-Booth, Y. (1971) Effects of brief separation from mother on rhesus monkeys, *Science*, **173**, pp. 111–118.

Petrie, M., Halliday, T. and Sanders, C. (1991) Peahens prefer peacocks with elaborate trains, *Animal Behaviour*, **41**, pp. 323–331.

Silver, R. (1978) The parental behavior of ring doves, *American Scientist*, **66**, pp. 209–215.

Further reading

Halliday, T. R. (1980) *Sexual Strategy*, Oxford University Press.

Slater, P. J. B. (1978) *Sex Hormones and Behaviour*, Edward Arnold.

CHAPTER 10
LIVING IN GROUPS

10.1 Introduction

Many animals live in groups for at least part of their lives. There is a continuum from animals that live an almost completely solitary existence, through animals that live some of the time in groups, to animals that spend their whole lives within the same group of individuals. Where animals live only some of the time in groups, group living is often a seasonal event, like winter flocking in many small birds.

Most animals also show *social behaviour* to some extent, that is behaviour shown by two or more animals interacting with each other. All sexually reproducing species, for example, must exhibit social behaviour at some point in their lives if they are to reproduce! Moles are solitary animals defending (to the death if necessary) their subterranean tunnel systems against the intrusion of another mole. Yet in the spring, for a few weeks, male moles tunnel their way into as many female territories as they can. No one has described their courtship and mating but all females of reproductive age subsequently examined are pregnant so they obviously have a brief period in their lives when they indulge in social behaviour. Being mammals, female moles also suckle and care for their young, so between birth and weaning females also interact socially with their offspring. Honey bee (*Apis mellifera*) colonies are at the other extreme in the degree of sociability shown. Their hives contain one female—the queen—who spends her whole time laying eggs and who depends upon the many female workers to collect food, feed her and her developing young, and build and maintain the hive. There are also a few males, called drones, whose only contribution is to mate with the queen. The social organization of the colony is very complex and no honey bee, even a worker, can survive alone, without the other members of the hive.

As social behaviour involves directing a behaviour pattern at another conspecific, some degree of recognition must be involved. Social behaviour need not necessarily involve recognition of an individual *per se*, however, but simply as a member of the same species or group. Where an individual can recognize another individual uniquely, rather than just as a conspecific, there can be the additional dimension of a social *relationship*. A social relationship exists where individuals recognize each other and interact with one another over a period of time. Mating behaviour and parental behaviour shown towards young after hatching or birth often involve the formation of social relationships.

Conspecifics living in a group do not *necessarily* exhibit social behaviour. Among animal groups there is a continuum from simple aggregations to groups, like honey bees, showing a complex social organization. If, for example, many individual insects are attracted to the same place for some reason, they will inevitably form a group. Those of you who have experienced an ever increasing number of biting insects around you on a summer's evening will probably have realized that you were not the centre of a *social* gathering but were merely a food source. Each insect was reacting to *you*, rather than responding to conspecifics, so no social behaviour is involved.

There can be many advantages in simply aggregating with conspecifics. For example, woodlice (*Porcellio*) tend to lose water by evaporation and in a dry place will rapidly dehydrate and die. They keep moving until they find themselves in a dark, damp area (Chapter 2), so aggregations can form simply because a number of individuals end up in the same damp, dark crevice. Each woodlouse will continue to lose water by evaporation but will lose less if huddled together with other woodlice than it would if it were alone, in the same way that a bundle of washing loses water less quickly than the same washing spread out. The lost water will also make the crevice more humid. As far as we know there is no social interaction as such between the woodlice but each individual benefits from the aggregation.

Living in groups can have disadvantages as well as advantages, however, i.e. costs as well as benefits. For example, a group of animals may keep themselves warm by huddling together but they also increase the risk of contracting a contagious disease. Social relationships may also involve conflict as well as cooperation, as in sexual relationships (Chapter 9). For example, a group-living individual may have to compete with other group members for food or mates. Whether an animal lives in a group, or joins one, depends on the balance between costs and benefits and this can vary between individuals and over time. Thus a particular benefit relating, say, to food availability might be seasonal, so group formation might also be seasonal. Furthermore, all group members may not benefit from group living by an equal amount. Under certain circumstances, those that benefit least may leave the group. This chapter will look at the adaptive basis of group living and the conflict and cooperation typical of social groups. Finally, it will discuss the phenomenon of cultural transmission of behaviour, by which behaviour can be passed on from one generation to the next by non-genetic means.

10.2 Living in groups and its adaptive basis

10.2.1 Predation

In relation to predation the possible cost to an individual of being in a group is that the group may be more conspicuous than the individual would be if it were on its own. Set against this is the decreased probability of being the individual who gets caught by the predator, provided that the predator does not eat all the group. If for example a particular predator always catches a prey animal once it has spotted it, then a solitary prey animal once seen by the predator will inevitably be caught, but the same individual prey animal in a group of ten has only a one in ten chance of being caught. This is known as the *dilution effect*. Thus the greater conspicuousness of a group of, say, ten individuals is only a net cost to an individual if a predator is more than ten times as likely to spot the group as to spot the lone animal.

Another benefit of group living may result from the group's ability to defend itself collectively. Defence has two aspects, *vigilance* and attack (including *mobbing* behaviour). Not all individuals will see a predator at the same moment, especially if they are occupied with feeding, for example. The more individuals there are to 'keep an eye open', the sooner the predator is likely to be spotted and, if the detection of a predator evokes a response (e.g. escape or an alarm call, see Section 4.3.7) then the rest of the group will get to know about the predator sooner than they would if they were alone. Thus group vigilance is greater than individual vigilance. Once a predator has been located individuals may flee from it or collectively attack it—behaviour less dangerous for a group than for individuals. Mobbing refers to the behaviour of some birds towards a predator, in which the birds fly around and at the predator together, calling loudly, until the predator is driven off.

These ideas about the possible costs and benefits of group living in relation to predation lead to predictions of how animals should behave in groups compared with when they are alone. They can also be tested experimentally.

The study of egg-shell removal by black-headed gulls (Section 4.3) showed that conspicuousness increases the chance of predation. Evidence that aggregations are more conspicuous than individuals comes from a study by Malte Andersson and Christer Wicklund (1978) of the fieldfare (*Turdus pilaris*), in the forests of Scandinavia.

Most birds nesting in forests choose nest sites that are some distance from neighbours and are well hidden, but the fieldfare is an exception. Its voluminous nests are made in deciduous trees and egg laying is complete before the trees are in leaf. Furthermore, the nests are most often found in colonies. Andersson and Wicklund made artificial fieldfare nests and put quail's eggs (which are similar to fieldfare eggs) in them. Nests were 'solitary' or arranged in 'colonies'. The artificial colonies were more heavily predated than were solitary nests, as shown in Figure 10.1.

In a 6-year study of natural nests, however, solitary fieldfare nests produced no fledglings whilst those in colonies produced, on average, 1.7 fledglings per nest.

☐   In the experiment all the eggs from 'colonial' artificial nests had been predated after 4 days, and yet in real colonies nests produce, on average, 1.7 fledglings. To what factor would you attribute this difference?

■   Adult birds are present at the natural nests, but absent at the artificial nests and the difference may be due to some feature of the adults' behaviour.

In fact, if you observe a predator (such as a crow) entering a colony you will see it being mobbed and defaecated upon by the fieldfares, and artificial nests placed near real colonies survived better than those placed near solitary fieldfare nests. Thus, in this species it would seem that the cost of group nesting—increased conspicuousness—is outweighed by the benefit of group defence.

A number of studies have shown that mobbing can reduce the success of predators. For example, kestrels (*Falco tinnunculus*) are mobbed at all times of the year by other birds and mobbing is effective in moving the kestrels away from the area; kestrels fly significantly further between foraging positions when they are mobbed than when they are not mobbed.

Greater vigilance has also been shown to reduce the success of predators. If many eyes are scanning then the chances are that the predator will be spotted earlier

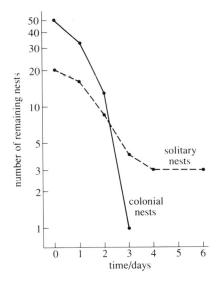

Figure 10.1   The pattern of predation at 'solitary' and 'colonial' artificial nests. (The colonial nest remaining on day 3 disappeared on day 4.) Predation is significantly higher at colonial nests. Note: The scale showing the number of remaining nests is not a linear scale, that is, each division is not equal and does not represent an equal number of nests. A scale such as this one is called a logarithmic scale (or log scale). It enables the graph to show what happens when there are very few nests left without being enormously tall.

than it would be by a solitary animal. Goshawks (*Accipiter gentilis*) are less successful when they attack large flocks of wood pigeons (*Columba palumbus*) and this is mainly because larger flocks take to the air when the goshawk is further away, i.e. the hawk is spotted earlier. Brian Bertram (1980) working in Africa obtained more direct evidence of the importance of group vigilance by noting the amount of time ostriches (*Struthio camelus*) spent looking around them when alone and when in flocks (see Figure 10.2).

Bertram's observations suggested that ostriches raised their heads at random time intervals and that each individual raised its head independently of what the others were doing. Figure 10.2 shows that overall, vigilance of the group is higher than it is for an individual, but nevertheless for over half the time no ostrich had its head up. Ostriches forage in long grass and they cannot see what is going on around them when their heads are down. Bertram checked this by waving at the group when all their heads were down! They raise their heads to scan their surroundings, on average, every minute or so, which does not give a predator, such as a lion, much time to stalk them and take them unawares. Thus, by being in a group the ostrich gains the benefits of increased vigilance whilst actually spending less time itself on scanning, therefore having more time available for other activities such as feeding.

This effect of increased group vigilance has been found in many species both when feeding and when resting. It might be thought that it would be possible for an individual to reduce its own vigilance—giving it more time to spend on feeding for example—and make use of the vigilance of other group members. Individual Thomson's gazelles (*Gazella thomsoni*) that are less vigilant are, however, more likely to be selected by cheetahs (*Acinonyx jubatus*) and are, therefore, more likely to be predated. When moving toward a group of gazelles a stalking cheetah has ample time to assess the vigilance of members of the group and chases the less vigilant of the two nearest individuals more often than would be expected by chance. It is also more likely to catch that animal than a more vigilant one. Thomson's gazelles do not always give an alarm call once they see a predator but they adopt a 'stare posture' and display a conspicuous white tail patch. Other gazelles in the group may be alerted only when they notice this behaviour, so scanning can serve to monitor the activity of other group members as well as to observe for predators directly. Individual gazelles vary in this behaviour and this gives some a better chance of survival than others.

Vigilance may also enable individuals to observe other group members in other contexts. Squirrel monkeys (*Saimiri sciureus*) and tamarins (*Saguinus labiatus*) are both equally 'vigilant', as measured by the percentage of time they break off from foraging to look around, but the squirrel monkeys, which have complex social relationships involving dominance hierarchies (see Section 10.3), spend 70% of that looking-around time apparently monitoring the behaviour of conspecifics whilst tamarins, which live in less hierarchical groups, spend less than 10% of their looking-around time monitoring one another's behaviour.

The dilution effect of being in a group may not be equally beneficial to all group members. It is commonly observed that the individuals on the edge of a group are most readily predated. This is true for Thomson's gazelles, for example. It is an advantage therefore to be at the centre of the group. There is a similar advantage

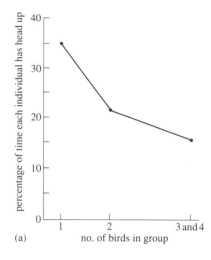

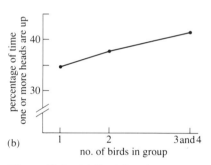

Figure 10.2 (a) Ostriches that are alone spend more of their time scanning for predators. (b) Overall vigilance increases as the group size increases.

to being at the centre of a nesting colony site and there is evidence that these places are often more keenly contested. W. D. Hamilton (1971) developed the idea of the *selfish herd*, in which any individual that can get to the centre of a group has a better chance of avoiding being eaten than those on the periphery, provided the predator does not eat all the group. Figure 10.3a shows a hypothetical pond with frogs arranged around the edge, equally spaced. Each frog has round it a large *domain of danger*. This is the area in which, should a predator such as a snake appear, the frog will be nearer to the snake than any of its conspecifics, and so is the most likely individual to be eaten. The frog reduces its chance of being eaten if it can reduce its domain of danger. It can do this by sitting between two other frogs, as shown in Figure 10.3b.

In a group of white-fronted geese (*Anser albifrons*), those individuals that were the greatest distance from a neighbour (i.e. that had around themselves a large domain of danger) were vigilant for longest.

☐ Does this support Hamilton's selfish herd model?

■ Not directly, though it suggests that it might apply.

☐ What evidence is needed to confirm Hamilton's model?

■ Evidence that shows that animals with a large domain of danger are more likely to be predated than those with a small domain of danger, and that individuals on the edge of the group attempt to move into the centre.

Julia Parrish, working in America, looked at the behaviour of fish in groups (schools) to assess the safest location within the school. Black seabass (*Centropristis striata*) attacking schools of Atlantic silversides (*Menidia menidia*) preferentially attack stragglers, which is in line with Hamilton's model. However, if there were no stragglers the seabass went for the centre of the shoal. Parrish makes the point that in the wild, fish schools are attacked by several different predators which may have differing modes of attack so the relative safety of a particular place in the school (or herd) may not remain constant. Furthermore, the position in the school may relate more directly to feeding opportunities, the silverside at the front of the school getting the best of the food. Thus the next section looks at the costs and benefits of group living in relation to feeding.

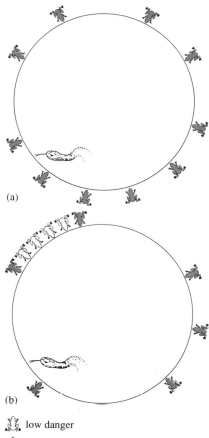

(a)

(b)

🐸 low danger

🐸 high danger

Figure 10.3 If a snake may strike at any point along the edge of the pond and always takes the nearest of the frogs sitting around the edge, all individuals are in high danger in (a) but the danger is lower for those which have jumped between others in (b).

10.2.2 Feeding

Weaver birds feeding on seeds in the savannah benefit from social feeding (Chapter 4) only because, when they find seeds, there are enough available to feed the whole flock. When a resource is limited there may not be enough for all the group and some may obtain less than others or may even go without.

Some experiments performed by Peter Major (1978) illustrate this clearly. He studied the jack (*Caranx ignobilis*), a predatory fish which forms temporary, small schools for activities such as feeding and reproduction, and its prey species the Hawaiian anchovy (*Stolephorus purpureus*), which is always found in schools. His study area was a lagoon in Hawaii that had been partitioned into enclosures to enable him to collect accurate data on predation.

☐ Study Figure 10.4a. On average, does the jack fare better when it hunts alone or when it is in a group?

■ The number of prey captured per individual is greater when the predators are working in groups.

☐ Study Figure 10.4b. Do all individuals benefit equally?

■ No. The leader does considerably better than those at the back of the group.

This illustrates a point made earlier, namely that group living may not benefit all members of the group equally.

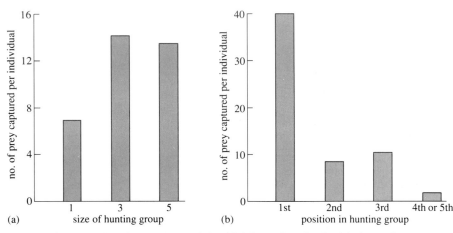

Figure 10.4 The jack is a predatory fish which hunts in schools: (a) shows the average number of prey captured per individual in relation to hunting group size, and (b) shows how the average number of prey captured per individual relates to that individual's position within the hunting group.

☐ Are the fish at the back of the group capturing more individuals than they would if they hunted alone? (Compare the success of the fourth and fifth fish in Figure 10.4b with that of the solitary fish in Figure 10.4a.)

■ No.

So why did the fourth and fifth fish join the school? Peter Major observed that the jack showed opportunistic behaviour on an individual basis. Any individual might be leader by detecting and orienting toward a school of anchovies. Others joined in as followers but in some cases overtook the leader. On average, each individual probably derives greater benefits than costs from joining a school as it will be in the lead on at least some occasions. Also, as with the black seabass, if there are no stragglers to snap up then the jack schools swim into the anchovy schools, breaking them up and capturing isolated anchovies before they can rejoin their school. Breaking up the anchovy schools is achieved more effectively by a group of jack than by a lone jack. In most cases, too, there is likely to be an abundance of prey items, so all the jack may be satiated even though there is initial competition for prey.

It should not be adaptive for an individual to reveal the whereabouts of food to others unless it also gains some benefit from doing so. However, it may be impossible to prevent other individuals from following it to a food source, or discovering about a food source from it. It is a common observation that feeding birds will be joined by others (e.g. at a bird table) but a number of investigators have examined the possibility that individuals may actually communicate information about a distant food source. It has long been known that honey bees can communicate information about the distance and direction of a food source by means of a 'waggle dance' (Book 3, Chapter 2). They benefit from this because it stimulates other bees to collect food for communal food stores. It has been suggested that bird roosts may similarly act as 'information centres' and Peter DeGroot (1980) tested for this possibility using wild-caught weaver birds. He had two groups, each of six birds. One group (A) had received a period of training and knew the whereabouts of a food source outside the communal aviary. The second group (B) had a similar training period but their knowledge was of the whereabouts of a water supply. Each group, therefore, was knowledgeable about one resource but naive about the other. Then, after an overnight period without food, Group B followed Group A to the food source, i.e. the birds that were naive for the resource in question followed the knowledgeable birds. After an overnight period without water, positions were reversed and Group A followed Group B to the water, naive birds again following knowledgeable birds. This occurs despite the fact that there are four possible exits from the aviary, that once a bird has left it cannot return, and that those birds still in the aviary cannot see out and so do not know whether departing birds have made a correct choice. He checked to see whether information was imparted overnight by allowing all birds to roost together but removing all knowledgeable birds at dawn. The naive birds did not now go to the right area for the resource of which they had been deprived. This shows that the birds can learn to find the resource they require when in the presence of knowledgeable birds. DeGroot also argued that in the wild, where roosts can contain thousands of birds, all birds would probably return to the roost at dusk with knowledge of some food resource, but some birds would have found a better food supply than others. It would be adaptive for those individuals who had only found poor feeding if they could follow other more fortunate individuals to better sites. To test this, DeGroot took two groups. One was taught the whereabouts of a chamber containing a feeding tray of pure seed, the other was taught the whereabouts of a chamber containing a tray of seed mixed with sand. Both groups were housed overnight in the same aviary and, the morning after, all birds went to the chamber with the pure seed. It is not known how the birds achieve this. One suggestion is that birds may reveal what they know by their appearance, knowledgeable birds moving in a purposeful way, whilst naive birds or those who only know of poor resources may reveal this by nervous behaviour, e.g. jerky movements and raised crest feathers.

Apart from the honey bee, in the examples given so far there has been no evidence that information has been deliberately communicated rather than just gleaned. However, Mark Elgar (1986) carried out an interesting study which showed that the behaviour of an individual which finds food may depend on how easy it is to share that food. He put out bread for house sparrows (*Passer domesticus*) either as crumbs or in a single slice. The feeder was on a roof top in Cambridge which was surrounded by a low wall. Birds could not see the feeder until they were on the wall. If they went down to the feeder they could not be seen by other birds. House sparrows give a chirrup call which Elgar found, by playing tapes of the call, attracted other sparrows.

☐   Is this sufficient evidence that only the chirrup call attracts sparrows?

■   No. It could be that any noise would have this effect.

☐   How could you eliminate this possibility?

■   You could play some other sound.

Elgar found that if the food was in crumbs the pioneer sparrow (i.e. the first one at the site) chirruped at a higher rate than if the same amount of food was in a slice. This is shown in Figure 10.5.

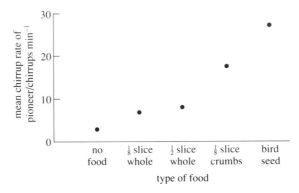

Figure 10.5   The mean chirrup rate of pioneers increased when the resources were more divisible.

☐   Look at Figure 10.5: why, do you suppose, that Elgar put out half a slice of bread as well as one-eighth of a slice of bread?

■   The half slice contains more actual food. The fact that the birds do not chirrup at a higher rate when there is a half slice of bread available shows that they are responding to the state of the food (divisible or indivisible) rather than the amount of food available.

It may seem strange that sparrows should attract others to a food source at all but it had previously been found that sparrows spend less time scanning for predators when they are in larger groups. This is a point not considered explicitly so far, namely that a behaviour may have a benefit in terms of one activity (here, reducing the risk of predation) whilst there is a cost in relation to another activity (here, increased competition for food). Looking at interactions of this kind is more difficult but is obviously necessary as these are the conditions under which the animals are normally operating.

10.2.3 A study of the interactions between costs and benefits of group living: Caraco's model

Starvation and predation are the two main causes of death of small birds in winter. They risk starving to death because most types of food become less abundant in winter. Additionally, because it is colder, they need more food to maintain their body temperature. Most of their searching is visual and with shorter daylight hours time available for foraging is at a premium. Predation is an ever present risk but it becomes more acute in winter because predators are hungrier, for the same reasons as their prey. They, like the small birds, are in danger of starving. The small birds need to be more vigilant than ever to avoid being eaten and, because they cannot scan and eat at the same time, this means that they have less time for foraging.

Under these conditions the question is, when will it pay a bird to hunt in a group and when will it be better off on its own? Furthermore, what size of group is optimal and will the optimal flock size change under different conditions?

To try to answer these questions Tom Caraco (1979) developed a mathematical model (the details of which will not be given here). In order to do this he made some assumptions, as follows:

1   As the flock size increases the amount of time each individual spends scanning for predators will decrease. In other words the small birds would behave like Bertram's ostriches. This gives them more eating time when they are in a flock.

2   As the flock size increases the amount of time available for feeding is increased (see assumption 1). Thus when their energy requirements have been met the amount of time spent behaving aggressively towards other birds will increase. The reason for this assumption is that whilst the individual may have eaten enough for the moment, resources are scarce in winter so it will benefit the individual to try to protect the remaining resources. In other words, competition and fighting will increase as flock size increases.

3   The birds can only do one thing at a time: eat, scan or fight (see Section 7.6).

There are other complications, which Caraco actually included in his calculations. One is a point raised earlier, that the benefits are not shared equally by all group members. Subordinate birds will spend more time on the move as successive dominants push them off the richest feeding areas. The dominant birds, however, will give a higher priority to feeding than to fighting. Finally, for all birds scanning is the highest priority activity and vigilance is never compromised. The model also took into account the predator attack rate and the probability of avoiding predation. Taking all these factors into consideration, Caraco calculated optimal flock size, i.e. the group size that maximizes individual survival. Figure 10.6a (*overleaf*) shows a simplified version of Caraco's model which assumes that time is spent only fighting, feeding or scanning. Figure 10.6b shows the model proper.

☐   In what way does Figure 10.6a differ from Figure 10.6b?

■   The optimal group size is lower in Figure 10.6b.

Caraco and his associates (Caraco *et al.*, 1980) then made observations of winter flocks of yellow eyed juncos (*Junco phaeonotus*), feeding in a canyon in Arizona and recorded the amount of time spent scanning, feeding and fighting. The results are given in Table 10.1 (*overleaf*).

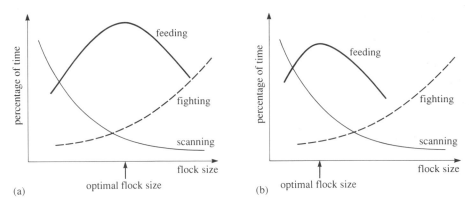

(a)

(b)

Figure 10.6 This shows models of optimal flock size: (a) is a simplified version showing fighting and scanning time at different flock densities and assumes all the remaining time can be used for feeding, and (b) shows the same data for fighting, feeding and scanning but in calculating the time available for feeding, predator attack rate and the probability of avoidance have also been included, as well as the different benefits available to dominant and subordinate birds.

Table 10.1 A test of Caraco's model's predictions. The percentage of time spent on feeding, scanning and fighting by different group sizes of yellow eyed juncos.

Flock size	Scanning	Feeding	Fighting
1 (n = 159)	30%	70%	0%
3–4 (n = 162)	13%	77%	10%
6–7 (n = 149)	7%	85%	8%

n is the sample size

Take the data from Table 10.1 and plot it onto Figure 10.7. When completed, compare your graph with Figure 10.8.

☐ Does this support Caraco's model?

■ Yes; in general, scanning changes in the direction predicted by the model.

☐ How does it differ from Figure 10.6?

■ Fighting is less markedly affected by flock size than in the model.

A number of predictions as to how different environmental factors will affect flock size were then made from this model. The first one concerns the amount of time spent fighting when the temperature increases or when food becomes more abundant.

At higher temperatures, or when food is more abundant, the birds will not need to spend so much time foraging.

☐ Why will higher temperatures mean that birds spend less time foraging?

■ They need less food to keep body temperature constant because they are not losing heat so quickly.

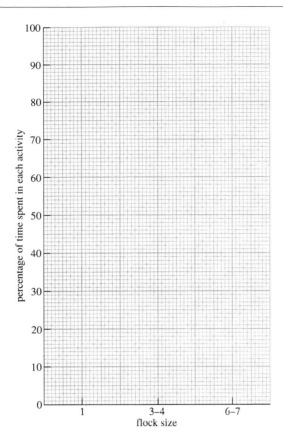

Figure 10.7   A blank graph for you to show how the proportion of time spent on three activities changes with flock size.

The model predicts that if they spend less time foraging they will have that time available for some other activity.

☐   Which activities could the time be spent on?

■   Scanning or fighting.

☐   Which activity will the model predict to increase?

■   Fighting. (Scanning you remember was assumed to be the highest priority activity so adequate time is already allotted to it.)

Thus the model predicts that at higher temperatures or when food supply is improved there will be a larger proportion of time spent on fighting and subordinates will leave the group (Prediction 1). This is a valuable prediction because it is not what one would intuitively expect, i.e. you might expect fighting to decrease if more food was available.

When measurements of behaviour and of flock size were made for yellow-eyed juncos at different temperatures, this prediction was found to be fulfilled. For example, the average flock size was seven birds at 2 °C but only two birds at

10 °C. At the higher temperature there was also an increase in the amount of fighting shown by the dominant birds.

The prediction was tested further by scattering extra food around the canyon. The result was the same: with more food available the birds were in smaller flocks.

The model also predicts that increased risk of predation will increase flock size (Prediction 2). Many models predict this; it is an extension of Hamilton's selfish herd model. However, peculiar to Caraco's model are the further predictions that scanning time will increase (the priority activity) and that this will take time away from something else.

☐ Which activities could decrease?

■ Feeding and fighting.

☐ Which one will decrease according to the model?

■ A certain food intake must be maintained. (Remember that the dominant birds give feeding priority over fighting.) Therefore, fighting will decrease.

Thus, increased risk of predation will increase flock size and scanning time, and decrease fighting (Prediction 2).

This was studied in two ways. A tame hawk was flown across the canyon to increase the risk of predation. On another occasion more cover was added to reduce predation risk by giving birds somewhere to hide. This was done by adding a small bush to the canyon! Table 10.2 shows the results of the experiment with the hawk.

Table 10.2   A test of the predictions of Caraco's model. The percentage of time spent on feeding, scanning and fighting by different group sizes of yellow eyed juncos when a hawk is flying above them.

Flock size	Scanning	Feeding	Fighting
1 (n = 79)	57%	43%	0%
3–4 (n = 52)	27%	69%	4%
6–7 (n = 162)	33%	60%	7%

n is the sample size

☐ Compare Table 10.1 and Table 10.2. Is there any evidence that flock size increases in the presence of a predator?

■ Yes. If you look at sample size you can see that proportionally far more birds are in flocks of six to seven birds. In fact the mean flock size was 3.9 birds without the hawk and 7.3 when the hawk was present.

☐ Use the data from Table 10.2 and plot it onto the graph paper given as Figure 10.7. Note how the proportion of time spent on the three activities changes with the increased risk of predation. Have the effects on scanning time and fighting changed in the direction predicted by the model?

■ Yes. The birds spend more time scanning and less time fighting.

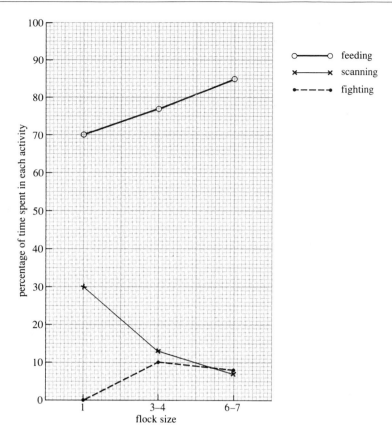

Figure 10.8   Data from Table 10.1.

Incidentally the birds near the bush also behaved as predicted. The effective predator risk was less for them and they did spend less time scanning and therefore more time feeding and fighting and the flock size decreased.

The model has thus been successful in increasing understanding of some of the selective forces which determine the flock size in small birds which feed in groups in the winter. It shows how different costs and benefits combine to affect group size.

☐   What was the mean flock size?

■   3.9 birds when there was no hawk around.

☐   Is this the optimal flock size?

■   No. In Figure 10.7 you will have shown that feeding time was greatest when there were six or seven birds in the group.

So why do birds not form a group of the optimum size so as to maximize their chances of survival? If birds are to maximize their feeding time they should be in larger flocks than those actually observed. These flocks are temporary aggregations of feeding birds, however: individuals come and go through the day. By looking at Figure 10.6 you can see why the optimal group size, for efficient foraging, might not be a stable group size. A lone bird arriving at the canyon would be better off joining a flock of optimal size rather than remaining alone or joining a smaller group. This is also true for the next bird to arrive, but eventually, as group size increases, the proportion of time spent feeding decreases (because of increased fighting according to this model) and birds would do better to leave this group and join a smaller group. What is actually observed is probably a stable group size rather than an optimal group size.

In nature, optimal group size will never exist because birds will always tend to join such flocks, making them larger.

10.3   Social conflict

As discussed in the previous section, living in a group confers both costs and benefits on individuals within that group. One particular kind of cost is the increased competition that arises when animals live in groups. If a solitary animal finds a choice food item it can eat it; if a group of individuals find it they have to compete for it. Within a group, there may also be a tendency for males to fight over sexually-receptive females. It is this increased competition in groups that explains, in part, why many species aggregate at one time of year but not at others. As described in Chapter 2, for example, starlings are highly sociable for most of the year, feeding and roosting in flocks, but in the spring they become highly aggressive and territorial (Figure 2.24). For starlings, competition for mates in the spring precludes group living.

Many social species show behaviour patterns that have the effect of reducing the incidence or severity of fighting. If a flock of hens that have just been brought together are put into a pen, they show frequent and often severe fights over food items and perch sites. After a few days, however, the level of fighting decreases markedly and close observation reveals that certain individuals in the flock are more likely than others to eat any item of food that is found, and to have priority of access to particular perches. The hens are said to have formed a 'pecking order', in which some individuals have established a status such that other birds quickly give way to them in any competitive context. Disputes are resolved quickly, often with only a brief display being involved, with one bird deferring quickly to the other. A social system in which certain individuals have priority of access to some resource, such as food or potential mates, is called a **dominance hierarchy**. If individual A regularly gains a resource at the expense of individual B, A is said to be *dominant* to B, and B is *subordinate* to A.

Dominance and subordinacy are relationships between pairs of individuals that result from learning. B becomes subordinate to A as a result of learning that it cannot defeat A in a fight. Thus, for a dominance hierarchy to be established in a group, individuals must have the capacity, first to recognise one another as individuals and, secondly, to learn from their previous encounters with those individuals. Consequently, dominance hierarchies are most apparent in social groups that remain stable, in terms of the individuals within them, over long periods of time.

In many species, dominance and subordinacy are correlated with morphological differences between individuals. Frequently, dominant individuals are simply larger than subordinates, but in some species there are features that serve as 'badges' of status. In the house sparrow (*Passer domesticus*), for example, the male has a black 'bib' extending from his beak down to the upper part of his chest. Dominant males have larger black patches than subordinate birds. There is considerable evidence that these 'badges' can act as signals of relative dominance, even between individuals that are not familiar with one another. They thus act as sign stimuli (see Chapter 2) that signal an individual's dominance status.

Dominance hierarchies are not necessarily static over long periods of time. Typically, younger individuals are subordinate to older ones but, as they grow older and become stronger, they may challenge their superiors, defeat them, and so rise up the hierarchy. Because a dominance hierarchy is built up from many pair-wise relationships, it will not always be linear; individual A may be dominant to B, and B dominant to C, but it is possible for C to be dominant over A. In groups with complex social organization, such as that found in some primates, individuals may form alliances in which two subordinate individuals act together to defeat an individual that is dominant over each of them. These subtleties are examined in the videotape *Social primates*.

10.4   Social cooperation

10.4.1   Altruism and kin selection

You learned in Chapter 4 that, in some species, individuals show behaviour that benefits others at a cost to their own individual fitness. Such altruistic behaviour is adaptive when the benefits in terms of the increase in the altruist's inclusive fitness are greater than the costs. This is most likely to be the case when the behaviour benefits the altruist's close relatives, as in alarm calling in Belding's ground squirrels.

There are many examples of cooperation in group-living species. In many bird species, for instance, a breeding pair are helped in rearing their offspring by other birds that do not breed themselves, even though they are sexually mature. What benefits could these helpers gain that would outweigh the reduction of their own reproductive success to zero? Glen Woolfenden and John Fitzpatrick (1984) have made a detailed study of one such bird, the Florida scrub jay (*Aphelocoma coerulescens*) (Figure 10.9, *overleaf*).

The Florida scrub jay lives in oak scrub. Pairs hold territories throughout the year, and just over half the pairs have helpers who contribute in two ways to raising the offspring of the breeding pair.

1   They help defend the young. This is done by giving alarm calls and mobbing predators such as snakes.

2   They bring food to the nestlings.

Table 10.3 shows that having helpers really does improve the breeding success of pairs.

Figure 10.9   The Florida scrub jay (*Aphelocoma coerulescens*).

Table 10.3   The average number of young fledged by experienced and inexperienced pairs with and without helpers.

	Number of young raised per pair	
	without help	with help
Inexperienced pair	1.24	2.20
Experienced pair	1.80	2.38

Woolfenden and Fitzpatrick found that helpers bring in about 30% of the nestlings' food, but that no more food overall comes into a nest that has helpers.

☐   Can you give the main reason for the improved survival of nestlings at nests with helpers?

■   Since, overall, no more food is provided, the main contribution that helpers make to nestling survival must be in their anti-predator behaviour.

☐   In what way does the breeding pair benefit from the helpers' food provisioning behaviour?

■   A pair with helpers has to do less food provisioning.

There is evidence that this improves their own physical condition. Their chance of surviving to the next year is 85% whereas it is 77% for pairs without helpers. This relates to a point made in Section 9.3.1, that there is a trade-off between reproductive and somatic effort. The effect of helpers is that a breeding pair can reduce their reproductive effort and so enhance their survival.

Thus there are clear benefits to the breeding pair and to the nestlings from this helping behaviour, but what do the helpers get out of it?

Most helpers are 1–2 years old and are physiologically capable of breeding. They are also almost always the offspring of the breeding pair. By increasing the annual survival rate of its parents the helper, on average, is making a small improvement to its own inclusive fitness. Put another way, the helper increases the chance that the parent is still around next year to produce more nestlings, and these nestlings will be siblings of the helpers.

Helping to rear siblings also increases the helpers' inclusive fitness. Suppose that there are two helpers helping every experienced pair. There is obviously a gain in inclusive fitness for each helper and this can be calculated.

☐   What is the gain, on average, in terms of the number of nestlings fledged by the experienced pairs with helpers?

■   Table 10.3 shows that without help they raise 1.8 young on average and this increases to 2.38 with help, so the average gain is 0.58 young.

☐   How much of this gain is attributed to each helper?

■   Half. There were two helpers so it follows that each helper has been responsible for the production of an average extra 0.29 of a nestling.

☐   If the helper had left home and bred what sort of productivity could be expected?

■   Table 10.3 shows that, on average, a pair of first time breeders produce 1.24 nestlings, so using the same logic as in the previous questions, each bird is responsible for 0.62 nestlings.

Even allowing for the fact that, by helping, the helper has slightly increased its inclusive fitness by increasing the probability that its parents survive an extra year, a productivity of 0.29 nestlings is a poor pay-off compared with 0.62 nestlings. A helper would appear to be better off breeding itself than staying and helping. So why does it do it?

Suppose we ask another question.

☐   Is the helper better off helping or not breeding?

■   Helping—a productivity of 0.29 nestlings is better than nothing.

This leads us to wonder whether there may be some other constraint on breeding.

Each breeding pair holds a feeding territory around its nest. The oak scrub is a limited habitat. The scrub is patchily distributed and there are large areas around each pocket of scrub that contain no suitable areas for breeding. There are not enough breeding territories for all the potential breeders, so it is not a choice between helping or breeding but a choice between helping or not breeding. In fact in the study from which this data has been taken there was a huge mortality rate from a mystery disease one year. Half the adult population died between 1979 and 1980.

☐ What would you predict to be the outcome of this?

■ If most nests had helpers and there were usually two helpers per nest then halving the adult population should mean that there would be enough territories available for most sexually mature birds to breed. (There would be some recruitment from the previous year's nestlings.) Most adults should breed and there would be few helpers.

This was indeed just what happened.

Thus kin selection partly accounts for the helping behaviour of the Florida scrub jay but environmental factors are also involved. There are ecological constraints on breeding and individuals are better off helping parents than they are not breeding.

The most extreme results of kin selection are seen in *eusocial* species. These are species such as the honey bee where there are sterile individuals who cooperate with the mother to bring up her young: like the Florida scrub jay helpers the sterile bees are rearing their siblings.

Eusociality appears in several species of insects, but the first time it was described in any other group was by Jennifer Jarvis in South Africa. Jarvis (1981) suggested that the naked mole rat (*Heterocephalus glaber*), a small mammal, was a eusocial species (see Figure 10.10). She had been studying their physiology and since 1974 had been trying to get females to come into oestrus (breeding condition) in captivity. It proved impossible to get more than one female in the colony to come into oestrus. Jarvis, together with the Americans Richard Alexander and Paul Sherman, then looked more closely at the colonial structure of this small mammal.

Figure 10.10   The naked mole rat (*Heterocephalus glaber*).

Mole rat colonies containing up to 100 individuals are found in the savannah of East Africa. More accurately they are under the savannah, living in burrows which are excavated and extended by workers who also make the nesting chambers and forage for the large tubers of the plant *Pyrenacantha kaurabassana* which are their main food. These tubers average 5 kg in weight, so the naked mole rats, averaging about 9 cm in length, bite off lumps to feed the queen and the larger workers, who stay near the queen and her young. Both males and females are workers but the females seem to be sterile. If the queen is removed one of the larger females can come into oestrus and take over as queen. The queen seems to suppress sexual behaviour in other females and in males except when she is in oestrus.

The main predators on the mole rat are snakes which can enter the burrow system. They may grab a small worker at times when groups are working near the surface, throwing out loose soil from new burrow systems. The alarm grunt of the victim stimulates the others to block the burrow in front of it; protecting the colony but leaving the small worker to its fate. If the danger continues the small workers flee toward the queen. She prods and nips the larger workers until they go out to attack the predator. Larger workers are more likely to be successful in driving off the predator. Two large workers in a captive colony, for example, were observed to bite a snake over and over again until it was dead.

☐   How could such a group of a breeding female plus sterile workers have evolved?

■   The members of the colony must be very closely related to each other.

This has been found to be so. Members of a colony are genetically so similar it is as though they had been inbreeding (see Chapter 3) for 60 generations. Ecological constraints may also have led to a colonial life. Colonies of naked mole rats occupy a habitat which is unsuitable for most mammals. The soil of the savannah is hard and dry and tubers are patchily distributed. A single mole rat could burrow for days without finding food, and even a pair of naked mole rats could probably not survive on their own.

10.4.2   Altruism towards non-relatives

The examples in the previous section showed that altruistic behaviour between relations can actually be cost-effective. However, altruistic acts toward non relatives have also been observed—how can we explain this? There seem to be two possibilities—*manipulation* and *mutualism*. Manipulation is the situation where the performer of the altruistic behaviour is duped by the recipient. An example of this is the altruistic behaviour of the host bird towards a cuckoo egg or chick in its nest. The host is duped because the egg mimics that of the host and is not recognized as 'counterfeit', and the chick presents supernormal stimuli. The spread of such a method of rearing young is limited—if the host species fails to rear any of its own young there will soon be no hosts!

Mutualism, or reciprocity, is the situation where individuals both perform and receive altruistic behaviour. If the benefit to the recipient is greater than the cost to the performer and if there are sufficient opportunities to reciprocate, both will accrue greater benefits than costs in the long term.

Mutualism has been demonstrated in vampire bats (*Desmodus rotundus*), which share a day-time roost and in which those that have not had a successful night foraging beg for food from their more fortunate roost-mates. G. S. Wilkinson (1984) captured eight bats as they left the roost and released them back into the colony at dawn without being fed. Of the eight bats, five were fed by other bats when they begged. All bats in the colony were individually marked and Wilkinson found that bats only fed those to whom they were related or with whom they frequently roosted.

Vampire bats cannot last very long without food, as Figure 10.11 shows. A bat feeds at night and spends the day in the roost, emerging at dusk to forage again. If it does not find food in the following 50 hours or so, it will die of starvation. Its weight loss after feeding follows the curve shown in Figure 10.11 where the 'pre-fed weight' is the weight of a bat at dusk following a previously successful night. Death occurs at about 75% of pre-fed weight.

☐ If a bat which has fed the previous night and is at point A on the graph gives 5% of its pre-fed weight to another individual how much closer (in hours) will it bring itself to starvation?

■ About $5\frac{1}{2}$ hours.

☐ If the recipient has been two nights without food and is at point B on the graph when it receives this donation how many hours farther away from starvation will it now be?

■ About 18 hours.

Thus the cost to the donor is less than the gain to the recipient. If you did not get these answers, look at Figure 10.12, which shows how they were obtained.

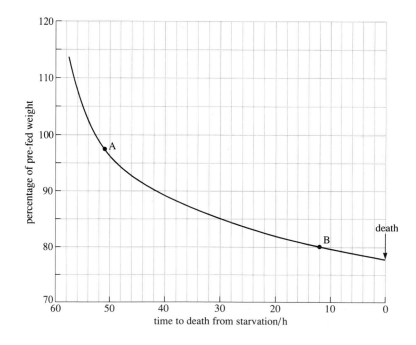

Figure 10.11 Graph showing weight loss after feeding as a percentage of the pre-fed weight of the vampire bat at dusk. Death occurs at 75% of the pre-fed weight at dusk.

The bats appear to recognise individuals in that they do not feed others indiscriminately. Bats share the same roost each night so there is plenty of opportunity for reciprocity. The cost to the donor of $5\frac{1}{2}$ hours (in the above example) is small compared with the gain by the recipient of 18 hours.

☐ Using the figures in the above example what is the net gain to each individual if 4 days later and under identical conditions the recipient becomes the donor and vice versa?

■ Net gain to each is $12\frac{1}{2}$ hours $(18 - 5\frac{1}{2}$ hours).

No other example of reciprocal altruism has been so amenable to quantification. There is however some doubt as to whether even this is an example of reciprocity rather than an example of kin selection.

10.4.3   Altruism that does not benefit the performer

The female Antarctic plunderfish (*Harpafiger bispinis*) guards her eggs. If she is removed, either by natural causes or by an experiment, then her place is taken by another fish (generally male). If this fish is removed another one takes over, and so on. The replacing fish are not related to the original female or to each other and there is no obvious gain to them. At present, there is no explanation for this behaviour. It may just be that not enough is known about this fish or what might be underlying this behaviour, or it could be that the behaviour is maladaptive.

It is often assumed that all behaviour will have a function, that it will have survival value or contribute to reproductive success. It is possible, however, that a particular behaviour might *not* have a function. For example, in the case of the antarctic plunderfish, pleiotropy might be involved, so that guarding behaviour is a consequence of some other behaviour just as people with PKU tend to have pale skin and fair hair as well as behavioural disorders (see Chapter 3). Another possibility is that the animal is currently inhabiting a different environment from the one in which the behaviour evolved.

For example, the little tern (*Sterna albifrons*) nests on sandy beaches in dispersed colonies. Its colonies are not dense enough to make colonial mobbing of predators effective. In Britain, its eggs are often predated or washed away by the tide, because the most suitable sandy beaches have been taken over by holiday makers and the birds are forced to nest in sub-optimal habitat.

10.5   Cultural transmission of behaviour

This chapter has so far made the implicit assumption that a behaviour pattern which is advantageous will be the result of the individual who performs this behaviour having a particular set of genes. Those offspring who inherit that particular set of genes will then also show the adaptive behaviour which will enhance their fitness. The inheritance of *acquired* characteristics is not normally possible by genetic means. If you learn that a particular berry is poisonous you will not eat it, but your children will not be born knowing that they should avoid eating that berry. The necessary information cannot be passed on to them genetically. It can however be passed to them in another way. They can be told to avoid it. In a similar way information can be passed between members of other species.

☐ What examples of this have you come across already?

■ Crows will not eat poisoned food after other crows have been poisoned (Section 6.3.5). Rats will learn to avoid noxious food if other rats in the group become ill after eating it (Section 6.4.1). Sparrows come to a feeder in response to calls from another sparrow (Section 10.2.2). Honey bees learn about food sources from other members of their group (Section 10.2.2).

Thus a learned response which is adaptive can sometimes be transmitted to other members of the group and to offspring. This kind of transmission of information by non-genetic means is known as *cultural transmission*. For example, shepherds training young sheepdogs in the presence of a trained sheepdog only need to provide the necessary stimuli: the adult dog 'trains' the puppy to make the appropriate responses.

Another well-known example of cultural transmission of information is the spread of the rather annoying habit that blue tits have of pecking through the foil of milk bottle tops and helping themselves to the creamy milk. This habit has spread on a geographical basis in a way that suggests that the birds were copying from others.

Another eating habit whose spread has been well documented is a particular food-preparation behaviour shown by the Japanese macaque (*Macaca fuscata*). These animals were being studied, in the wild, by a group of Japanese scientists. It is an irritating fact well known to ethologists that when you are watching animals in the wild they perform all the really interesting behaviour when they are just out of your view. To try to combat this the scientists supplied the monkeys with sweet potatoes, laid out on the beach beside the lake, an area in full view of the observers. Before long, one of the macaques, a 16-month-old female called Imo, was seen washing her potatoes in a nearby stream. This efficient method of removing the sand was soon copied by others, particularly those of her own age. Within 10 years the habit had spread to all those in the population, except for the very young, less than a year old, and those over 12 years of age.

Imo was also the initiator of another food preparation habit. Grains of rice were also scattered on the beach. The macaques at first picked the grains out of the sand, singly. One day, when she was 3 years old Imo was seen to scoop a handful of sand and grain and throw the whole lot into the water. The sand sinks and the rice floats. The rice can then be scooped off the surface of the water. Imo did not necessarily have the intelligent foresight to work out what would happen. However, having thrown the mixture of sand and grain into the water (for whatever reason) and observed the outcome she had the ability to recognize the consequences of her behaviour and to appreciate the increased efficiency of foraging in this way. Others in the group learnt from her. Again, those of her own age copied her first, then mothers learned from their young, with the adult males being the last group to learn the technique.

Behaviour may change in response to the environment. As shown above, such adaptive changes can sometimes be transmitted culturally, the equivalent of being 'inherited' by non-genetic means.

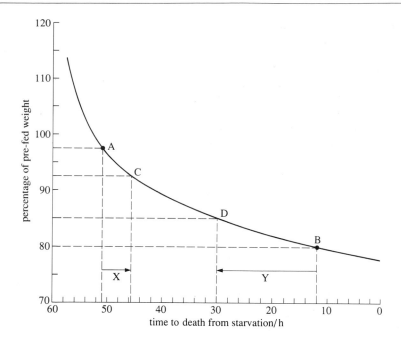

Figure 10.12   Graph showing the cost to a donor and the benefit to a recipient of a blood meal. At point A on the graph, the donor makes the donation when at a pre-fed weight of 97.5%. A 5% pre-fed weight donation takes it to point C on the graph, so the number of hours lost is the distance along the horizontal axis from point A to point C, i.e. distance X, which is about $51 - 45\frac{1}{2} = 5\frac{1}{2}$ hours. The recipient has been two nights without food and, at point B on the graph, is at 80% of its pre-fed weight and only about 12 hours away from death by starvation. A donation of 5% of pre-fed weight takes it back to 85% pre-fed weight, point D on the graph, where it is 30 hours away from starvation. Thus the gain in hours is the distance along the horizontal axis from point B to point D, i.e. distance Y, which is about $30 - 12$ hours $= 18$ hours.

Suppose, however, that potatoes had not been supplied by the scientists so that the first time that potatoes were washed was not observed, and instead potato washing was a behaviour pattern shown by all macaques from the time that the species was first discovered. Observers might then have assumed it to be an evolved, genetic trait. Without carrying out the appropriate experiments, biologists have no way of knowing how many supposedly genetically-based behavioural characteristics are actually culturally acquired.

Summary of Chapter 10

There is a continuum from animals that are solitary to those that live always in a group. There is also a continuum in the amount and complexity of social behaviour shown by different species. Where individuals join and leave groups it is possible to study the costs and benefits of their behaviour by making appropriate observations and by conducting experiments.

The adaptive basis of living in groups was considered in relation to predation and feeding. The cost of conspicuousness can be off-set by the benefits of group defence and group vigilance. The cost of competition can be off-set by more effective foraging behaviour. In reality, an animal's behaviour can be very variable and the costs and benefits of a particular activity will vary depending on circumstances. Caraco's model of the time spent on vigilance, feeding and fighting led to predictions about how group size might change as environmental conditions changed. These predictions were borne out in studies of natural populations.

Conflict within groups is a cost of group-living but, in stable groups, this is often minimized by the formation of dominance hierarchies. The evolution of social cooperation can be understood as a result of the processes of natural selection as long as the net benefits of cooperation outweigh the costs. This can be true for altruistic behaviour which is reciprocated or which is shown towards kin. Altruistic behaviour where the net benefits appear to be less than the net costs cannot as yet be explained but the species may not be at a stable stage in its evolution.

A learned adaptive change in behaviour can spread throughout a population as a result of cultural transmission. Unless the learning process is observed, however, it is impossible to tell, without carrying out appropriate genetic experiments, whether a particular behaviour pattern has spread culturally or by the process of natural selection operating on genetically-based characteristics.

Objectives for Chapter 10

When you have completed this chapter you should be able to:

10.1   Define and use, or recognise definitions and applications of each of the terms printed in **bold** in the text.

10.2   Distinguish between social behaviour and a social relationship. (*Question 10.3*)

10.3   Describe and give examples of the costs and benefits of living in groups. (*Questions 10.1 and 10.2*)

10.4   Understand Caraco's model and the predictions that follow from it. (*Question 10.2*)

10.5   Interpret a model which is presented in graphical form and make simple predictions from it. (*Question 10.2*)

10.6   Explain what is meant by a dominance hierarchy and describe the effect that it has on the amount of aggressive behaviour that occurs in social groups. (*Question 10.1*)

10.7   Define altruistic behaviour and show that in most cases it is explicable in terms of natural selection.

10.8   Give examples of altruistic behaviour that are not explicable in terms of natural selection and suggest reasons for their occurrence.

10.9   Explain how adaptive behaviour can be transmitted by non-genetic means.

Questions for Chapter 10

Question 10.1 (*Objectives 10.3 and 10.6*)

A particular carnivore commonly hunts in groups of five. When a group kills a prey animal, group members share it unequally, with the most dominant animal getting the most, and the most subordinate animal the least. If several prey are killed the most dominant animal will become satiated and eat progressively less, and so on down the hierarchy, until all individuals are satiated. If there is a shortage of prey items the most subordinate animal may go hungry. Suggest possible reasons why it does not leave the group and hunt alone, either temporarily or permanently.

Question 10.2 (*Objectives 10.3, 10.4 and 10.5*)

Caraco's model predicts that a flock size of six to seven birds would be optimal because foraging time is greatest (Table 10.1). Section 10.2.3 suggested one reason why this optimal size might not be stable. From your study of Caraco's data (Tables 10.1 and 10.2) can you suggest any other reason?

Question 10.3 (*Objective 10.2*)

Figure 10.13 presents data on two human mothers during breast-feeding interactions with their babies. Which of the data provide information solely about the interactions of the mother and infant and which also give information about their relationships? Data for one mother and baby are denoted by A, data for the other by B.

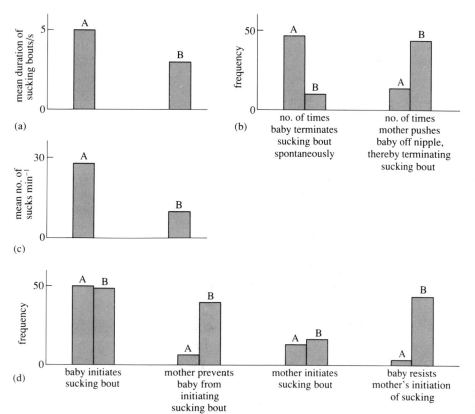

Figure 10.13   Data on two human mothers during breast-feeding interactions with their babies.

References

Andersson, M. and Wicklund, C. G. (1978) Clumping versus spacing out: experiments on nest predation in fieldfares (*Turdus pilaris*), *Animal Behaviour*, **26**, pp. 1207–1212.

Bertram, B. C. R. (1980) Vigilance and group size in ostriches, *Animal Behaviour*, **28**, pp. 278–286.

Caraco, T. (1979) Time budgeting and group size: a theory, *Ecology*, **60**, pp. 611–617.

Caraco, T. (1979) Time budgeting and group size: a test of theory, *Ecology*, **60**, pp. 618–627.

Caraco, T., Martindale, S. and Pulliam, H. R. (1980) Avian flocking in the presence of a predator, *Nature*, **285**, pp. 400–401.

DeGroot, P. (1980) Information transfer in a socially roosting weaver bird (*Quelea quelea*: Ploceinae): an experimental study, *Animal Behaviour*, **28**, pp. 1249–1254.

Elgar, M. A. (1986) House sparrows establish foraging flocks by giving chirrup calls if the resources are divisible, *Animal Behaviour*, **34**, pp. 169–174.

Hamilton, W. D. (1971) Geometry for the selfish herd, *Journal of Theoretical Biology*, **31**, pp. 295–311.

Jarvis, J. U. M. (1981) Eusociality in a mammal: cooperative breeding in naked mole-rat colonies, *Science*, **212**, pp. 571–573.

Major, P. (1978) Predator-prey interactions in two schooling fishes, *Caranx ignobilis* and *Stolephorus purpureus*, *Animal Behaviour*, **26**, pp. 760–777.

Wilkinson, G. S. (1984) Reciprocal food sharing in vampire bats, *Nature*, **309**, pp. 181–184.

Woolfenden, G. E. and Fitzpatrick, J. W. (1984) *The Florida Scrub Jay: Demographic Attributes of a Cooperative Breeder*, Princeton University Press.

Further reading

Trivers, R. (1985) *Social Evolution*, Benjamin/Cummings Publishing Co.

Wilson, E. O. (1980) *Sociobiology: the Abridged Edition*, Harvard University Press.

GENERAL FURTHER READING

Alcock, J. (1997) *Animal Behaviour: An Evolutionary Approach,* 6th edn, Sinauer Associates Inc.

Krebs, J. R. and Davies, N. B. (1993) *An Introduction to Behavioural Ecology*, 3rd edn, Blackwell Science Ltd.

McFarland, D. J. (1993) *Animal Behaviour: Psychobiology, Ethology and Evolution*, 2nd edn, Addison Wesley Longman Ltd, Longman Resources Unit, Harlow.

Manning, A. and Dawkins, M. S. (1998) *An Introduction to Animal Behaviour*, 5th edn, Cambridge University Press.

Ridley, M. (1986) *Animal Behaviour: A Concise Introduction*, Blackwell Scientific Publications, Oxford.

Slater, P. J. B. and Halliday, T. R. (eds) (1994) *Behaviour and Evolution*, Cambridge University Press.

ANSWERS TO QUESTIONS

Chapter 1

Question 1.1

Statement (i) refers to (c) the function of birdsong, statement (ii) refers to (a) its causation, statement (iii) refers to (b) its development. Statement (iv) tells us something about (a) the causation of behaviour, but also gives us insight into (c) its function.

Chapter 2

Question 2.1

Exafferent information comes from visually-perceived changes in the position of the insect that result from movements of the insect itself. Reafferent information comes from visually-perceived changes in the position of the insect that result from the lizard's own movements. The lizard must separate the reafferent information from the exafferent information to determine the exact location of the insect.

Question 2.2

(i) Exteroceptive; (ii) enteroceptive; (iii) proprioceptive.

Question 2.3

(a) Both (i) and (ii) are consistent with the FAP concept. (b) (i) is consistent, but (ii) is not; clearly learning is involved. (c) (i) is consistent, but (ii) is not; if stereotypies vary between individuals, they cannot be typical of a species. (d) Both (i) and (ii) are consistent with the FAP concept.

Question 2.4

You would need to do an experiment, in which you separated the effects of light and the song of other birds. One approach could be to isolate an individual bird from the song of other birds and see whether it still sang at dawn. Alternatively, you could keep the test bird in the dark and let it hear the dawn chorus.

Question 2.5

(a) False; a sign stimulus is a particular feature of an egg that elicits retrieval. (In this example, an egg provides more than one sign stimulus.) (b) True. (c) False; a supernormal stimulus is one that has dimensions falling *outside* the normal range of variation in the context in which that stimulus is normally relevant. Supernormal stimuli may be artificial (e.g. very large dummy eggs) or natural, e.g. a cuckoo's egg laid in the nest of another species.

Question 2.6

(a) A releaser is an external stimulus that, in Lorenz's scheme, elicits a fixed action pattern. (b) A stimulus–response chain is a sequence of behaviour, involving two animals, in which a response performed by one animal acts as a stimulus for a response by the partner. (c) A consummatory stimulus is an external stimulus that causes an animal to cease what it has been doing and switch to something else. (d) A taxis is a pattern of movement which is directed with respect to the stimulus that elicits that movement. (e) A kinesis is a pattern of movement which is not directed with respect to the source of stimulus that elicits that movement. (f) A pheromone is a chemical substance produced by an individual animal that alters the behaviour or physiology of another individual. (g) A hormone is a chemical compound that is secreted within the body and that has various effects on the behaviour and physiology of an animal.

Question 2.7

Statement (a) suggests that stimulus filtering in female frogs is central, statement (b) that it is peripheral.

Chapter 3

Question 3.1

A locus is a particular place on a chromosome; a locus is occupied by a particular gene which produces a particular protein; a gene can come in a number of different forms called alleles; if the identical allele is on both chromosomes of a chromosome pair then that allele is homozygous; the genotype is the whole constellation of alleles in a particular organism.

Question 3.2

Sentence (a)   A string of DNA bases is a gene or an allele (provided, of course, that it is a long enough string).

Sentence (b)   A string of DNA bases codes for a protein.

Question 3.3

The concept in the statement that is correct is that chromosomes are found in gametes and two gametes combine to produce a zygote. Thus, chromosomes are passed from one generation to the next via the gametes. However, the chromosomes in the gamete are different from the chromosomes that the parent had, because the process of crossing over produces novel combinations of alleles in each chromosome.

Question 3.4

A given living organism can only have a maximum of two forms of a single gene (i.e. two alleles); one form on each of a pair of chromosomes. But there may be three, four or more alleles, some of which might be found in other conspecifics. However, the real answer would have to include all potential alleles, not just those actually found. Most alleles result in a completely defective protein which is lethal to the organism. And the number of these lethal alleles? Virtually infinite.

Question 3.5

Tfm results in tissue that cannot detect testosterone. One of the consequences of this is that the organism is less aggressive. The normal gene produces the protein that detects testosterone; it does *not* produce aggression.

Question 3.6

This is an example of pleiotropy.

Question 3.7

You could do a selective breeding experiment where you carefully observe prey killing in a number of cats and allow to mate together only those cats with a similar prey killing style. If this was repeated over several generations, and there was a genetic component to prey killing, then strains with distinct prey killing styles should emerge.

A second possibility is to examine the prey killing of different strains of cat.

A third possibility would be to examine prey killing in closely related species.

The last two would require the examination of prey killing in the offspring of crosses and back-crosses.

Question 3.8

1   The particular allele associated with striking at slugs may be recessive and absent in the non-slug-eating population. The offspring of the cross would be heterozygous and so the effects of the recessive allele would not show up.

2   The effect of the 'slug' allele may be masked by other genes from the 'non-slug' parent.

3   The laboratory environment may have been unsuitable for the expression of the 'slug' gene.

Chapter 4

Question 4.1

1 and 5 are polymorphisms because they are distinct types found within the same population. 2 and 3 are neither because they refer to variation *between* and not within species. 4 is an example of polytypic variation because distinct types are found in different parts of the species' range.

Question 4.2

(a) True. (b) False. Natural selection has no 'purpose' as such and, although the relative fitnesses of members of the population may change as a result of selection, the species as a whole does not necessarily 'improve'. (c) False. In a constant environment, any selection pressures will also be constant but natural selection need not be non-existent. (d) False. Stabilizing selection does not lead to evolution. (e) True. Evolution can result from chance factors as well as from natural selection. (f) False. Natural selection is the process of differential lifetime reproductive success and survival is only one factor contributing to reproductive success. (g) False. Natural selection can favour altruism if it enables relatives to increase their reproductive success; it cannot normally favour altruism towards non-relatives even if it enables the group in general to benefit.

Question 4.3

(a) Lifetime reproductive success and longevity. (b) (i)–(iii) are all components of fitness since they all contribute to lifetime reproductive success.

Question 4.4

(a) Relative fitness is 1.0 because this pair produced most offspring. Selection coefficient is $1.0 - 1.0 = 0$. (b) This pair produced half as many offspring (3) as the most successful pair (6), so their relative fitness is $3/6 = 0.5$. Selection coefficient is $1.0 - 0.5 = 0.5$. (c) The average number of offspring produced in the population $= (2 + 6 + 1 + 0 + 3 + 0 + 4 + 0 + 1 + 3)/10 = 20/10 = 2$. Relative fitness of an average pair is $2/6 = 0.33$. Selection coefficient is $1.0 - 0.33 = 0.67$.

Question 4.5

(a) Directional selection—average tail-raising rate increases and range of variation changes. (b) Disruptive selection—individuals showing average tail-raising rates are selected against, individuals showing extreme tail-raising rates are favoured.

Question 4.6

Reproductive success of the young adults is reduced because they are not rearing their own offspring. Young adults might benefit if they are related to the adults they help and if, as a result of their altruistic behaviour, the breeding birds rear more offspring. The young adults are more likely to benefit if the possibility of rearing their own offspring is small—perhaps, for example, because only older, more experienced birds are able to obtain breeding territories.

Question 4.7

r is 0.125. There is a 25% chance that two half siblings will both inherit a copy of a specific allele carried by one or other of their parents. (Because they have only one parent in common, they are only half as likely to inherit the same specific allele as are full siblings, with both parents in common.) A child has a 50% chance of inheriting a copy of a specific allele carried by its parent. Thus there is only a 50% of 25% chance, i.e. a 12.5% chance (or a probability of $0.5 \times 0.25 = 0.125$) that an aunt and a half-niece will both carry copies of the same specific allele.

Question 4.8

(a) Costs: use of energy that could be used in e.g. mating or feeding; possible injury, thereby jeopardizing future reproductive success; loss of potential mate to a satellite. Benefits: it prevents his territory (or a female) being taken over by the other male, so he continues mating with females attracted to his territory. Benefits are likely to outweigh costs because the resource fought over (effectively access to mates) is very valuable. So the behaviour is likely to be adaptive. (b) Costs: energy put into making milk instead of improving her own condition and thereby improving her own future reproductive success; time spent in suckling when she could be feeding herself. Benefits: offspring is given a better start in life, improving its chances of surviving and producing grandchildren. Since the mother is unable to have a new offspring that year, she can probably improve her reproductive success by passing on any 'spare' energy to her previous offspring. So the behaviour is probably adaptive.

Question 4.9

1   Studying natural variation in behaviour within a species;

2   comparing closely related species that occupy different habitats and relating differences in behaviour to differences in habitat;

3   comparing closely related species in similar habitats that are presumed to have been faced with similar selection pressures to see if they evolve similar behaviour; and

4   an experiment.

Chapter 5

Question 5.1

The sensitive period is a discrete period of time when an organism can be influenced by a particular stimulus. Any one organism would have many, overlapping, sensitive periods, each for a particular stimulus or set of stimuli.

Question 5.2

The general form of an organism can be predicted, but not the detail; not the height, weight, hair colour, aptitude, friendliness or aggressiveness. The notion of an absolute blueprint would suggest that these are predictable elements.

Question 5.3

Amongst other possibilities, the mother could influence the temperature of the nest, and the amount and intensity of contact with the pups (e.g. licking and nursing). The physiology of the mothers of the two strains might also differ so that they produce milk in different quantities and of different composition.

Question 5.4

In filial imprinting a young animal imprints on the visual characteristics of its mother.

In sexual imprinting a young animal learns the characteristics of an adult or sibling and subsequently chooses a mate with similar characteristics.

In olfactory imprinting a young animal acquires a preference for a particular smell.

Question 5.5

The evidence for the song template is based mainly on the finding that, isolated from an early age, birds sing a species-specific rudimentary song. As such birds have never heard this species-specific song, it must be generated internally, from a template.

Chapter 6

Question 6.1

The puppy could be injected with sex hormones, such as testosterone, to simulate adult levels of sex hormone. In fact, when this is done the puppy urinates in the same manner as the adult, so the change in behaviour is the result of sexual maturation rather than learning.

Question 6.2

When rats are satiated with food, put them in a complex environment containing food but with wire mesh to prevent them from eating or attempting to eat. Let them explore it for, say 30 minutes each day for a week. Make them hungry, put them back with the wire mesh removed, and see how easily they get to the food location. Compare this with the behaviour of rats familiar with the same environment for the same length of time but with no food present during familiarisation. If the first group are better at finding the food (which in practice they are), then learning occurred during the 30-minute sessions, and only later was revealed in performance. (Incidentally, this is known as *latent learning*.)

Question 6.3

The rat should normally discriminate food on the basis of taste and smell, whereas birds should normally discriminate on the basis of visual cues.

Question 6.4

There is an important difference. In the Pavlovian, or classical conditioning situation, the bell and food must be closely associated in time; the food must come either at the same time as the bell or just after. Even in the taste aversion experiment (Figure 6.13), the illness must follow the food and a specific association is formed between nausea and the food. In sensitization, the two stimuli do not necessarily occur close together in time. Furthermore, not just the original response (e.g. gill withdrawal) but also other responses are sensitized. (However, the similarity between the two processes—that the response to one stimulus is a function of the response to another—is important, and, in an evolutionary sense, sensitization may possibly be a primitive precursor of the sort of learning that is brought about by classical conditioning.)

Question 6.5

(a) The rat's startle response is a short-lived response. It is completed well before 10 seconds have elapsed. The gill-withdrawal response shown by *Aplysia* is also rapid but it takes a while for the gill to be re-extended and the stimulus in this case cannot be re-presented after 10 seconds because it takes longer than this for the gill to recover fully from the previous stimulus.

(b) Habituation is described as a simple form of learning that is not retained for long periods, and spontaneous recovery is full recovery of the original response. However, incomplete recovery of the response after periods of medium length (30 minutes, 60 minutes) shows that these animals have retained some memory of the habituation.

Chapter 7

Question 7.1

There are a number of possible sequences. (a) Female rats are sometimes receptive and motivated and at other times uninterested, as a function of their oestrous cycle. Imagine that, at the time in question, a female has come into the phase of her cycle in which she is motivated. (b) She might press a lever in a Skinner box to earn the reward of access to a male rat, thereby showing purposive behaviour. (c) When she is sexually motivated and engaged in mating, other activities that would normally occupy her time such as feeding and drinking are temporarily inhibited. (d) Finally, prior to mounting by the male she shows the species-typical pattern of lordosis that facilitates penile insertion.

Question 7.2

When first placed in the maze, the animal's feeding motivation is very low and it does not ingest any food. However, it learns the location of food. Subsequently, when put in the maze after a period of food deprivation, it has a strong feeding motivation and the prior learning is revealed in its finding the food.

Question 7.3

Objectively, as more food is ingested, the person might eat less energetically, show more discrimination and be more easily lured into other activities. Subjectively, the pleasure rating given to food might decrease and even become negative.

Question 7.4

(a) By the criterion of inhibiting other activities, defence and aggression are unambiguously candidates. All other candidates for behaviour, e.g. mating, feeding or exploring, can be inhibited and thereby suspended while the animal performs defensive behaviour. (b) Animals can recruit a variety of different strategies to put a distance between themselves and any danger. (c) The freezing behaviour of a frightened rat is a species-typical response potentiated by the motivational state. Only the criterion of changing responsiveness presents any problems with such a classification. The tendency to show defensive and aggressive behaviour patterns may show little fluctuation over periods of time, as might be expected in some circumstances from adaptive considerations. However, in some species (e.g. red deer) one can see a distinct fluctuation in intraspecific aggression as a function of testosterone level.

Question 7.5

Not really. There is no identifiable physiological variable that is held within limits by sexual behaviour. No physiological harm is caused by sexual abstinence, in spite of often-heard claims to the contrary! This does not mean to say that there is no physiological basis to sex—indeed there must be something within the organism, e.g. within the nervous system, that fluctuates and thereby leads to fluctuations in sexual behaviour.

Question 7.6

(a) A low energy state promotes feeding and ingested food then corrects the low energy state, i.e. *decreasing* energy levels lead to *increased* feeding. (b) Similarly, a *decrease* in the level of body fluids below normal *increases* drinking motivation. Ingested water then restores body fluids to normal. (c) A deviation in body temperature in either direction from $37\,°C$ motivates behavioural temperature regulation which tends to correct the deviation. A *decrease* in temperature increases the motivation of the animal to find a warm place and thus *increases* its temperature, while an *increase* in temperature decreases it motivation to find a warm place and thus *decreases* its temperature. In all three cases an *increase* in one factor leads to a *decrease* in another factor which in turn affects the first factor, hence the term negative feedback.

Question 7.7

By injecting concentrated salt solution, water is moved from the cells into the extracellular space (the blood). This deficit stimulates drinking. In the intact animal, water passing the mouth and in the stomach exerts inhibition upon drinking motivation and this will play a major role in terminating drinking. If these pathways are cut or blocked, the animal should drink much more than normal.

Question 7.8

An example of feedforward is given by drinking in association with a meal. A rat ingesting dry food drinks too rapidly after the meal for drinking to be explained by movement of water from blood to gut. Rather, the rat drinks in anticipation of such movement and thereby pre-empts it, giving an example of feedforward.

Positive feedback is probably evident at the start of drinking. Drinking at first increases drinking motivation, an example of positive feedback. Drinking also has a negative feedback effect. Some time after the start of drinking, the increasing negative effect will outweigh the positive effect and drinking motivation will decrease.

Question 7.9

Group A would be expected to press the lever since their salt-deficient state will have created a motivation to obtain salt. In time they will give up when no salt appears. Group B would be expected to have no motivation to press the lever and would not do so, or at least would do so very little.

Question 7.10

Food pulls water from the body fluids into the gut, thus disturbing body fluids. Therefore reducing food intake reduces this disturbance. A compromise is established between the conflicting demands of normal food ingestion, which would disturb body fluids, and no food ingestion which would be bad for energy regulation. It is assumed that the compromise maximizes the animal's fitness.

Chapter 8

Question 8.1
The pupil light reflex is simply the fixed and quite automatic consequence of an increase in light intensity. Similarly, the knee-jerk reflex is a fixed and inflexible consequence of mechanical stimulation. There is no uncertainty about their appearance, except in diseased conditions. Therefore the term 'cognitive', which refers specifically to processes involving flexibility and choice, is inappropriate.

Question 8.2
The hormone corticosterone responds, with an elevation in secretion rate to above the baseline level, to both extinction and the receipt of a reward less than expected. Animals tend to slow up their rate of running for reward in response to both situations.

Question 8.3
(a) It would be expected to turn towards goal box 1, which is indeed what rats tend to do in this situation. (b) It would follow a procedural rule such as 'always turn left at a choice point' and so would turn towards goal box 2.

Question 8.4
The rat familiarizes itself with the alternative routes and with their current state of repair. This knowledge could prove useful in the future when the straight-ahead route might be blocked. The rat would be expected to take route B if there were a block in the straight-ahead route that could be avoided by taking this detour. It would be expected to take C if the block were on the straight-ahead pathway between the exits from routes B and C.

Question 8.5
One might argue that the rat is able to perceive the platform visually. By adding milk, Morris ruled out the possibility of directly sensing the location of the platform.

Question 8.6
You could test the vigour of their bar-pressing under extinction conditions in the two Skinner boxes. If the vigour of pressing in box X is increased then this would be evidence that the information is held in a declarative form. Two bits of information would be held: (1) food A will correct the vitamin deficiency, and (2) pressing the lever in Skinner box X is followed by access to food A. Were the information to be held as a procedural rule e.g. 'eat food A' or 'press lever' then the rat would not be expected to show increased responding under extinction conditions in box X relative to box Y.

Question 8.7
Animal A might have observed that on every occasion when a warning signal X had been made by animal B there really was danger present. Signal X might be the appropriate warning of a snake creeping up from below. Then animal B emits a warning signal of different quality Y, appropriate for a swooping eagle. This is

the first time that monkey A has heard B emit this warning signal. A takes action appropriate to there being an eagle present, *abstracting* the reliability of B's information.

Question 8.8

It rules out, for example, the case that seeing a strange monkey with a red mark causes an individual to touch its ear for some reason unconnected with self-recognition. It might, for example, evoke intense fear and intense fear might cause the animal to touch its ears.

Chapter 9

Question 9.1

(a) A short-day species is one in which individuals come into reproductive condition in response to shortening daylength in autumn. (b) Anisogamy is the existence within a species of two types of gamete, e.g. eggs and sperm, that differ in size and, usually, behaviour. (c) Polygyny is a mating system in which individual males mate with more than one female, but females mate with only one male.

Question 9.2

(a) Ultimate factors relate to breeding occurring at the optimal time: calves are born when food is most abundant. (b) Proximate factors relate to environmental cues that elicit breeding activity: mating is a response to shortening daylength.

Question 9.3

As sperm are very small, and therefore each requires less energy to produce than eggs, they are produced in much larger numbers. The result is that males have a greater reproductive potential than females and it is possible for males to increase their reproductive success by seeking multiple matings to a greater extent than females.

Question 9.4

The overall abundance and distribution of the food supply can promote polygyny. Rich food supplies make it more likely that one parent will be able to rear the young on its own. Patchy food supplies make it more likely that individual males can control food resources essential to females and thus control access to females.

Question 9.5

No. Strictly speaking, polygyny is a property of the individual. In the pied flycatcher, some males are polygynous, some are monogamous, and some have no mates.

Question 9.6

A transgenerational effect is one in which the development of behaviour in one generation is affected by experiential or learning effects, as opposed to genetic effects, in the previous generation.

Chapter 10

Question 10.1

It may not be capable of capturing sufficient food alone, i.e. although it does not always get enough food with the group, it gets more than it would alone. It may be exposed to predators on its own, i.e. the cost of not getting enough food might be outweighed by the benefit of lower predation risk. Being of low status may also be a temporary phenomenon; the individual will gain greater benefits as it moves up the dominance hierarchy. Finally, it may not be possible for it to go off on its own to hunt because, if it detects a prey item and tries to do so, the rest of the group may notice and follow it.

Question 10.2

When a predator is overhead the smaller groups (three to four birds) are spending a greater proportion of time feeding than are the birds in the larger group. Thus this flock size is optimal when predators are around. In the natural environment, these kinds of environmental change occur rapidly and changes in flock size may lag behind.

Question 10.3

(a) depicts the duration of sucking bouts, and (c) depicts their vigour. These are properties of interactions. (b) and (d) also show interactions, such as mother initiating a suckling bout, but taken together, they also provide information about the relationship between mother and baby in pair A compared with pair B. On the basis of the measures shown in (b) and (d), mother and baby A seem to have a more harmonious relationship than mother and baby B.

GLOSSARY

adaptive In the evolutionary sense of the term, a character is adaptive when the total fitness benefits associated with it outweigh the total fitness costs. (Section 4.4)

adrenalin Hormone, secreted mainly by the adrenal glands situated next to the kidneys, in response to stressful situations. Also called epinephrine, e.g. in the USA. (Section 2.9)

allele Any form of a gene that can occur at the locus occupied by that gene. (Section 3.2.3)

altricial Requiring a period of growth and development after birth or hatching before becoming fully mobile, unlike precocial animals. (Section 5.1)

altruism In the evolutionary sense of the term, the performing of behaviour that benefits another individual at a cost to the performer's own individual fitness. (Section 4.3.7)

amino acid A molecule which, when it joins together with other amino acids, forms a protein; about twenty different amino acids are found in living organisms. (Section 3.2.2)

anisogamy The existence of gametes in two distinct forms (in animals, large, usually immobile eggs and small, usually mobile sperm). (Section 9.3.1)

associative learning Learning as a result of exposing animals to two or more stimuli that have a particular relationship to one another. The animal demonstrates associative learning if it shows evidence of recognizing the relationship between the two stimuli. Classical conditioning and operant conditioning are examples of associative learning. (Section 6.3)

back-cross A mating between an offspring and an individual of one of the parental strains. (Section 3.3.1)

base A type of molecule which, combined with phosphate and sugar molecules, forms DNA. There are four types of base and the order they occur in DNA determines the order of amino acids in the protein for which that section of DNA provides the genetic code. (Section 3.2.2)

causation In the ethological sense of the term, aspects of behaviour that are analysed in terms of responses to specific objects and events. The specific objects and/or events that precede a behaviour and that cause that behaviour to occur. (Section 1.1.1)

character Any aspect of an organism's behaviour, physiology or morphology. All the characters of a particular organism together make up its phenotype. (Section 3.2.3)

chromosome A thread-like structure consisting of DNA and protein usually found in the nucleus of the cell. (Section 3.2.3)

classical conditioning   A procedure for producing behavioural modification via associative learning. In classical conditioning a stimulus, not normally associated with a particular response (a neutral stimulus), comes to be associated with a stimulus (the unconditional stimulus) that *is* normally associated with that response, so that it can elicit a response even in the absence of the unconditional stimulus. (Section 6.3.2)

coefficient of relatedness (*r*)   The probability that a particular allele in one individual will be inherited from a common ancestor by another individual. (Section 4.3.7)

coevolution   Evolution occurring simultaneously in different species, or different sexes, in which each becomes adapted as the result of the selection pressures each species, or sex, imposes on the other. (Section 2.8)

cognitive map   Information relating to the spatial properties of the environment stored in memory in such a way as to preserve the spatial relationship between features within that environment. (Section 8.2.1)

competition   A situation in which the demand for a resource is greater than the supply available, so that some or all individuals get less than they need. See *direct competition*, *indirect competition*. (Section 4.3.2)

complex learning   Any example of learning that appears to be more complex than associative learning. It includes imprinting, song-learning, insight, the ability to reason, and cognition. (Section 6.3.5)

component of fitness   A character that contributes to overall fitness. It may increase or decrease fitness. (Section 4.3.4)

conditional response (CR)   The term used to describe a response which, after classical conditioning, is elicited by the conditional stimulus in the absence of the unconditional stimulus. (Section 6.3.3)

conditional stimulus (CS)   The term used to describe the neutral stimulus once it has become associated with the unconditional stimulus in classical conditioning, i.e. once it is able to elicit a response (the conditional response) in the absence of the normal stimulus for that response. (Section 6.3.3)

conspecific   An individual of the same species. (Section 2.8)

consummatory stimulus   An object or event, related to a pattern of behaviour and the outcome of it, which causes an animal to cease performing that pattern of behaviour. (Section 2.6.1)

continuous variation   A character such as height, size, colour or metabolic rate, that shows continuous variation may take any value whatsoever, between certain limits. Using such a character, individuals cannot be divided into clearly distinguishable categories, between which there are no intermediates. (Section 4.2.1)

Coolidge effect   An increase in sexual motivation as a result of changing to a new mating partner. (Section 7.3)

courtship   Male and female behaviour patterns that precede and accompany the act of mating. (Section 1.2.1)

crossing over The process that occurs during meiosis where pairs of chromosomes overlap and exchange matching sections. See *recombination*. (Section 3.2.4)

declarative representation A form of information stored in memory which involves knowledge about the world but which does not involve instructions on what to do based upon that knowledge. (Section 8.2.2)

deoxyribonucleic acid See *DNA*. (Section 3.2.2)

deprivation experiment An experiment in which animals are reared under conditions in which they are deprived of some particular aspect of their normal developmental environment. (Section 1.2.3)

development Changes in an organism that occur during its life as a result of growth, differentiation, maturation and/or experience. (Sections 1.1.1 and 5.2)

differentiation The process in development by which unspecialized cells become specialized to perform a particular role. (Section 5.2)

direct competition Direct physical interaction with competitors, such as physically blocking other individuals' access to a resource or fighting with them to obtain it. (Section 4.3.2)

directional selection A type of natural selection that acts in such a way that the proportions of different alleles or phenotypes in a particular population consistently increase or decrease over a number of generations. (Section 4.3.6)

discrete variation (or discontinuous variation) A character shows discrete variation when individuals can be divided into clearly distinguishable categories, between which there are no intermediates. The human blood groups, A, B, O, etc., are an example of discrete variation. (Section 4.2.1)

dishabituation The recovery of a habituated response after receiving a significant novel stimulus. (Section 6.3.1)

display A pattern of behaviour, often very stereotyped, that has a communication function. Most commonly applied to postures and other visual signals, but can also be used for auditory or olfactory signals. (Section 1.2.1)

disruptive selection A type of natural selection that acts in such a way that the proportions of extreme phenotypes, or the alleles associated with them, increase consistently over a number of generations while the proportions of average phenotypes, or the alleles associated with them, decrease. (Section 4.3.6)

DNA The molecule deoxyribonucleic acid that carries the genetic information of the cell, coded in the sequence of bases; the main constituent of a chromosome. (Section 3.2.2)

dominance hierarchy A social system in which certain individuals have priority of access to some resource, such as food or potential mates. A ranking of individuals in a group in order of the number of other individuals each dominates, so that the highest ranking dominates all others in the group and the lowest ranking dominates none of the others. (Section 10.3)

dominant If an animal is heterozygous at a particular locus, and if the phenotypic character of only one of the alleles at that locus is shown in the phenotype, then that allele is said to be dominant to the other allele. Dominance is relative, not absolute, i.e. allele A may be dominant to allele B but recessive to allele C. It may also be complete or partial (Section 3.2.3). The term is also used to describe the social relationship between two individuals: if one individual is dominant to another then it has priority of access to some resource such as food or potential mates. See *dominance hierarchy*. (Section 10.2.3)

endocrine gland   An organ or group of cells that produces a hormone which is secreted internally, i.e. into the blood. (Section 2.9)

enteroceptor   A receptor that responds to a change in the internal conditions within an animal's body. (Section 2.3.2)

environment   Everything other than the genotype, including not only the physical world in which an organism lives, and all the other living organisms with which it interacts directly or indirectly, but also the internal environment of the individual. (Section 4.2.1)

enzyme   A protein that controls the rate at which chemical reactions take place in a living organism; a biological catalyst. (Section 3.2.2)

ethogram   A detailed description of a related set of behaviour patterns. (Section 2.4.2)

ethology   The study of the behaviour of animals in their natural environment. (Section 1.1.1)

exafferent   With reference to the stimulation of an animal's sense organs as a consequence of changes in the animal's external environment that do not result from the animal's own movements. (Section 2.3)

exteroceptor   A receptor that responds to stimuli external to the animal. (Section 2.3.2)

extinction   The loss of a learned response. The cessation of a particular conditional response, either as a result of disruption of the association between a conditional stimulus and an unconditional stimulus (see *classical conditioning*) or by the removal of reinforcement for that behaviour (see *operant conditioning*). The conditional response is also said to have been extinguished. (Section 6.3.3)

extracellular   Outside the cell. (Section 7.2.2)

feedforward   A process whereby a potential disturbance to a homeostatic state is recognized and action taken to correct the disturbance before it actually happens. (Section 7.2.2)

fitness (or relative fitness)   A measure of lifetime reproductive success (LRS) relative to other members of a population, i.e. the LRS of the individual concerned divided by the LRS of the most successful member of the population. This can be expressed either as a proportion (i.e. a number between zero and one) or as a percentage. (Section 4.3.4)

fixed   An allele is *fixed* in a particular population when it is the only allele found at a particular locus within that population. A phenotype may also be said to be fixed when it is the only phenotype present in a particular population (i.e. there is no polymorphism). (Section 4.3.5)

fixed action pattern (FAP)   A pattern of behaviour that is totally stereotyped in its performance, for all individuals in a species showing that behaviour pattern and on every occasion when it is performed. (Section 2.4.1)

function   The fitness benefits gained from the possession of a particular character. (Sections 1.1.1 and 2.2)

gamete   A sex cell with half the normal complement of chromosomes which, during sexual reproduction, will fuse with another gamete to form a zygote with the full complement of chromosomes. In animals the gametes are sperm and eggs. (Section 3.2.4)

gender   In the sense used in Book 1, the phenotypic sex of an individual. (Section 3.6.4)

gene   A length of DNA which carries the information necessary to produce a particular protein. (Section 3.2.3)

genetic code   The sequence of bases in the DNA that carries information about the correct position of amino acids relative to one another in proteins. (Section 3.2.2)

genetic drift   Chance fluctuations in allele frequency within a population. (Section 4.3.6)

genetic mosaic   An organism that has cells of two distinct genotypes. (Section 3.6.1)

genotype   All the alleles within a cell or carried by an organism. (Section 3.2.3)

gland   An organ or group of cells that secretes a particular substance or substances, e.g. hormones, enzymes. See *endocrine gland*. (Section 2.9)

group selection   A type of selection that is hypothesized to operate at the group level rather than at the individual level as it does in natural selection. In the process of group selection, groups favoured by selection have a higher rate of 'reproduction' (i.e. they split to form new groups that colonize new areas) than groups selected against, which die out completely and are replaced by colonizing groups of the type favoured by selection. (Section 4.3.7)

growth   An increase in size resulting from an increase in the number of cells in an organism and/or an increase in the size of those cells. (Section 5.2)

habituation   The loss of an unlearned response, i.e the effectiveness with which a stimulus elicits a response is reduced if the stimulus is repeatedly applied with no serious consequences to the organism. (Section 6.3.1)

heterogeneous summation   Phenomenon in which different aspects of a stimulus have an additive effect in eliciting a response. (Section 2.5.1)

heterozygous   The condition in which the two alleles at a particular locus on a pair of homologous (matching) chromosomes are different. (Section 3.2.3)

homeostasis   The maintenance of a stable state. In living organisms, the maintenance of a stable internal environment. (Section 7.2.2)

homozygous   The condition in which the two alleles at a particular locus on a pair of homologous (matching) chromosomes are the same. (Section 3.2.3)

hormone   A chemical stimulus secreted by an endocrine gland and transported in the blood stream to its site or sites of action. Examples include testosterone, oestrogen and adrenalin. (Section 2.9)

hybrid   The offspring of a mating between individuals belonging to two different types, e.g. two different species, sub-species or strains. (Section 3.5.1)

imitation learning   The ability to learn by copying others. (Section 6.3.5)

imprinting   The rapid learning process in which a young animal learns particular characteristics of its environment, usually the characteristics of its mother. (Section 5.5)

inclusive fitness   The lifetime reproductive success (LRS) of an individual carrying a particular allele, plus the LRSs of other individuals also carrying that allele as a result of descent from a common ancestor. (Section 4.3.7)

independent (or **random**) **assortment**   The process by which, during meiosis, each chromosome from a matching pair is randomly separated into one of two gametes independently of how other matching pairs of chromosomes are divided. (Section 3.2.4)

indirect competition   Competition that does not involve direct physical interaction with competitors, such as being more skilled at finding food or having a more efficient digestive system. (Section 4.3.2)

innate   Performed correctly the first time, apparently without any experience or practice. Innate behaviour is thus present or potentially present at birth or hatching, though it is not necessarily inherited, i.e. the possibility of environmentally influenced acquisition of a particular innate character between conception and birth or hatching is not excluded. (Section 1.2.3)

insight   The ability to perceive the solution to a problem without actually working through the intervening steps and without using trial and error. (Section 6.3.5)

instinctive drift   The situation in which a conditional response is not retained but instead gradually changes towards a response that is more common in the animal's behavioural repertoire, i.e there is a drift to the more natural species-typical behaviour. (Section 6.4.1)

instrumental conditioning   A procedure for producing behavioural modification via associative learning. See *operant conditioning*. (Section 6.3.2)

intracellular   Inside the cell. (Section 7.2.2)

kin selection   A type of natural selection operating indirectly on an allele by favouring the relatives of the individual having the allele, since these are likely also to be carrying the allele concerned by descent from a common ancestor. (Section 4.3.7)

kinesis (plural kineses)   Movement by an organism in response to a stimulus, which is not oriented with reference to the position of that stimulus. (Section 2.7.1)

learning   The process by which a memory of experience is acquired, often detected as a response by an animal to a novel situation such that, when confronted subsequently with a comparable situation, the animal's behaviour is reliably modified. (Section 6.1)

lifetime reproductive success   The number of fertile offspring an individual produces during its lifetime that survive to breed themselves. (Section 4.3.4)

locus (plural loci)   A place on a chromosome occupied by a single gene. (Section 3.2.3)

long-day species   A species that commences reproductive activity in response to the lengthening daylength that occurs in the spring. (Section 9.2.1)

maturation   A combination of growth and differentiation. (Section 5.2)

meiosis   A type of cell division that results in the halving of the number of chromosomes as a necessary prelude to the formation of gametes. (Section 3.2.4)

memory   A change in the nervous system consequent on learning, by which information is stored—the inferred intervening process that connects learning to recall. (Section 6.1)

mitosis   A type of cell division in which one cell divides to form two cells, each identical to the original cell. (Section 3.2.4)

modal action pattern (MAP)   A pattern of behaviour that is not totally stereotyped in its performance, but which shows a high degree of stereotypy, between different individuals and between different performances of the behaviour by the same individual. (Section 2.4.1)

monogamy   Condition in which individuals have one mating partner, at least over a certain time period such as one breeding season or one reproductive cycle. (Section 9.3.3)

motivational system   A system consisting of a particular motivation, and its associated physiology and behaviour. For example, the drinking motivational system consists of drinking motivation, drinking behaviour and the relevant physiology of the body, e.g. body-fluids and their control. (Section 7.1)

mutant   An individual carrying an allele that has undergone a mutation; the phenotype associated with a particular mutation. (Section 3.2.5)

mutation   An alteration in the order of bases in a DNA molecule. (Section 3.2.5)

natural selection   The process of differential survival and reproduction, by which the number of individuals possessing a particular allele or characteristic changes over the course of one or more generations. (Section 4.3.3)

negative feedback   Negative feedback is the principal process by which homeostasis is maintained. It operates by making a comparison between the current state of the system and the desired state: the difference between the two (the 'error') is then reduced by causing action to be taken that will move the system in the *opposite* direction to the error. The thermostat in a central-heating system operates by negative feedback. When the temperature rises above the desired level, the thermostat registers an error. The error is 'positive' because the current state is greater than the desired state. The thermostat therefore switches *off* the heater. When the temperature decreases below the desired level, the error is negative because the current state is less than the desired state. The thermostat therefore switches *on* the heater. (Section 7.2.2)

negative reinforcement   See *reinforcement*. (Section 6.3.4)

neutral mutation   A mutation that results in no measurable effect on the overall phenotype of the individual carrying the mutant allele, i.e. that has no measurable effect on the individual's fitness. (Section 3.2.5)

neutral stimulus (NS)   A stimulus not normally associated with a particular unconditional stimulus but which becomes associated with that UCS during classical conditioning. (Section 6.3.3)

non-associative learning   Learning brought about by habituation or sensitization. It occurs when an animal is repeatedly exposed to a single stimulus. (Section 6.3.1)

oestrogen   A hormone produced mainly by the ovaries, involved in the development and maintenance of many female sexual characteristics. (Section 2.9)

operant conditioning   A particular form of instrumental conditioning. The use of reinforcement (positive or negative) or punishment to select, strengthen or weaken a behaviour pattern (the operant) shown spontaneously by an animal. For example, a rat can be conditioned to press a lever in its cage by positively reinforcing this behaviour with food. The rat will show the operant of lever pressing more frequently as a result. (Section 6.3.2)

operant   See *operant conditioning*. (Section 6.3.4)

peripheral filtering   Stimulus filtering as a result of selectivity in responsiveness of part of the peripheral nervous system (i.e. those parts of the nervous system other than the brain and spinal cord—see Book 2). It may be based for example on the properties of the sense organs, which may respond only to certain types of stimuli. (Section 2.7.2)

phenocopy   An organism with the phenotype usually shown by a particular mutant but possessing the genotype of a 'normal' (or wild-type) non-mutant individual. (Section 3.6.4)

phenotype   Generally, the sum of all the characters that an organism possesses; the term can also be used specifically to refer to the expression of (i.e. the character produced by) a particular allele. (Section 3.2.3)

phenylketonuria   See *PKU*. (Section 3.3.3)

pheromone   A chemical substance produced by one individual that alters the behaviour and/or physiology of another individual. (Section 2.7.1)

phylogeny   The evolutionary history of an organism, i.e. the descent of species, individuals or characters from ancestral species, individuals or characters. (Section 1.1.2)

PKU   A potentially debilitating, genetically caused, human disease in which the amino acid phenylalanine cannot be broken down in the body, leading to mental retardation, for example, if untreated. (Section 3.3.3)

pleiotropy   The condition in which an allele has more than one effect on the phenotype, i.e. it affects more than one character. (Section 3.4)

polyandry   Condition in which a female has more than one mate within a single reproductive cycle or breeding season, but males have a maximum of one mate. (Section 9.3.3)

polygamy   Condition in which individuals have more than one mating partner within a single reproductive cycle or breeding season. (Section 9.3.3)

polygeny   The condition in which a character is the result of the combined effects of more than one gene. (Section 3.5)

polygyny   Condition in which a male has more than one mate within a single reproductive cycle or breeding season, but females have a maximum of one mate. (Section 9.3.3)

polymorphism   The situation in which a character is found in two or more discrete forms within a single population. (Section 4.2.1)

polytypy   The situation in which a character is found in two or more discrete forms, each in a different part of the species' range. (Section 4.2.2)

population   Any group of individuals from the same species and from a particular area, which may or may not have something else in common. For example, all the blackbirds in Nottingham or all the red-haired people living in Bradford. (Section 4.2.1)

positive feedback   A process which acts in the opposite way to negative feedback. It operates by making a comparison between the current state of the system and the desired state: the difference between the two (the·'error') is then *increased further* by causing action to be taken that will move the system in the *same* direction as the error. In the analogy of the central heating system used to illustrate negative feedback, the thermostat, if it operated under a positive feedback system, would, as the temperature rose, cause the heater to produce even more heat, sending the temperature even higher. (Section 7.2.2)

positive reinforcement   See *reinforcement*. (Section 6.3.4)

precocial   Fully mobile very soon after birth or hatching, unlike altricial animals. (Section 5.1)

procedural rule   A form of information stored in memory which tells the individual what to do in a given situation. (Section 8.2.2)

proprioceptor   A receptor that responds to the movements or position of the muscles in an animal's body. (Section 2.3.2)

protein   A molecule formed from a number of amino acids joined together in a string. (Section 3.2.2)

punishment   An aversive stimulus, such as a mild electric shock, given every time an animal performs a particular operant, which leads to that operant no longer being performed. Note that punishment is not the same as negative reinforcement, in which the animal has to perform a particular operant in order to *avoid* or *escape from* an aversive stimulus. (Section 6.3.4)

random assortment   See *independent assortment*. (Section 3.2.4)

range   The geographical area occupied by a particular species. (Section 4.2.2)

reafference principle The principle that an animal, if it is to respond appropriately to objects and events in its environment, must be able to differentiate between exafferent and reafferent sensory input. (Section 2.3)

reafferent With reference to stimulation of an animal's sense organs that results from changes in the animal's relationship with its external environment, not from changes in the environment itself. (Section 2.3)

reasoning The ability to follow through a series of abstract associations in order to arrive at a theoretical conclusion. (Section 6.3.5)

recall The successful use of learned experience; the expression of a response modified as the result of learning. Recall depends on the extent to which the material was originally stored in memory and the extent to which it is still available. (Section 6.1)

receptor A sensory cell which responds to a particular kind of stimulus, e.g. a light receptor responds to light. Some receptors, such as proprioceptors, respond to an internal stimulus; others, such as light receptors and touch receptors, respond to an external stimulus. (Section 2.3)

recessive If an animal is heterozygous at a particular locus, and if the phenotypic character of only one of the alleles at that locus is shown, then the allele whose effect is not shown is said to be recessive to the other allele. See *dominant*. (Section 3.2.3)

recombination The combined processes of crossing over and independent assortment. (Section 3.2.4)

reflex A totally stereotyped and specific response to a specific stimulus. (Section 2.2)

reinforcement A stimulus given by an experimenter to an experimental animal contingent upon the animal performing a particular behaviour pattern. Negative reinforcement encourages a particular behaviour pattern by making that behaviour pattern necessary if the animal is to avoid or escape from an unpleasant stimulus. Positive reinforcement encourages a particular behaviour pattern by the presentation of a pleasant stimulus such as food every time the desired behaviour pattern is performed. (Section 6.3.4)

relative fitness See *fitness*. (Section 4.3.4)

reproductive effort That part of the total energy that an organism has available that is expended on reproductive activities, as opposed to that expended on growth and survival (somatic effort). (Section 9.3.1)

response A pattern of behaviour that occurs as a consequence of some preceding event. (Section 2.1)

selection coefficient A measure of selection pressure, expressed as the difference between the relative fitness of the best phenotype (always 1.0) and the relative fitness of an inferior phenotype. (Section 4.3.4)

selection pressure The strength of selection for or against a particular character or phenotype. This can be expressed as the selection coefficient. (Section 4.3.4)

sense (or sensory modality)   A system by which stimuli of a particular type or consisting of a particular type of energy are received by an animal, e.g. the sense of vision receives light energy, the sense of hearing receives sound energy, the senses of taste and smell receive chemical stimuli and the sense of touch receives mechanical stimuli. (Section 2.3.2)

sensitive period   A specific period of time during an organism's development, when it shows an enhanced responsiveness to particular stimuli. (Sections 1.2.3 and 5.5.2)

sensitization   A form of learning in which responsiveness to a standard stimulus increases. (Section 6.3.1)

sensory modality   See *sense*. (Section 2.3.2)

sexual imprinting   The process by which a sexually immature animal learns the characteristics of those animals it will mate with when it reaches adulthood. (Section 5.5.5)

shaping   The process of reinforcing behaviour patterns which approximate more and more closely to a desired operant. (Section 6.3.4)

short-day species   A species that commences reproductive activity in response to the shortening daylength that occurs in the autumn. (Section 9.2.1)

sign stimulus   The particular feature of a complex stimulus that actually elicits the response. (Section 2.5)

somatic effort   That part of the total energy that an organism has available that is expended on non-reproductive activities, principally growth and survival. (Section 9.3.1)

sonagram   A graphical representation of a sound. (Section 3.5.1)

species   A group of individual organisms that can mate and produce fertile offspring. (Preface)

species-specific   A term applied to any character, including a behaviour pattern, that only occurs in one particular species. (Section 1.2.3)

spontaneous recovery   The recovery of a habituated response after the elapse of a period of time. (Section 6.3.1)

stabilizing selection   A type of natural selection that acts in such a way that the proportions of each allele or phenotype in a particular population remain constant. (Section 4.3.6)

stimulus (plural stimuli)   An object or event that an organism can detect and which may cause a particular event, such as a pattern of behaviour, to occur. (Section 2.1)

strain   A group of conspecifics which are genotypically and phenotypically similar to each other but genotypically and phenotypically different from other conspecifics not of the same strain. (Section 3.3.1)

supernormal stimulus   An object or event (usually but not always artificial) that elicits a stronger response than the natural stimulus normally associated with that response. (Section 2.5.1)

survival value The contribution made by any character, including a behaviour pattern, to an individual's survival. (Section 1.1.1)

taxis (plural taxes) Movement by an organism in response to a stimulus that is oriented with reference to the position of that stimulus. (Section 2.7.1)

territory A defended area. (Section 1.2.2)

testosterone A hormone, produced mainly by the testes, which is involved in the development and maintenance of many behavioural and physical characteristics, especially but not exclusively sexual characteristics, mainly in males but also in females. (Section 2.9)

unconditional response (UCR) The response normally elicited by the unconditional stimulus in classical conditioning. (Section 6.3.3)

unconditional stimulus (UCS) The natural stimulus of a particular response (the unconditional response) in classical conditioning. (Section 6.3.3)

wild-type The 'normal' phenotype of an organism; a non-mutant. (Section 3.3.2)

zygote The cell formed by the fusion of a male and a female gamete; in animals, a fertilized egg. (Section 3.2.1)

ACKNOWLEDGEMENTS

Grateful acknowledgement is made to the following sources for permission to reproduce material in this book:

FIGURES

Figure 1.2: courtesy of Dr Tim Halliday; *Figure 1.3:* Krebs, J.R. (1977) *Animal Behaviour*, **25**, pp. 475–478, Baillière Tindall; *Figures 2.1, 2.2a, 2.10, 2.18, 3.13, 4.5, 5.5, 7.13, 7.17:* McFarland, D. (1985) *Animal Behaviour*, Longman Group UK; *Figures 2.2b, 2.9, 2.11a, 2.12:* reprinted from Tinbergen, N. (1951) *The Study of Instinct*, by permission of Oxford University Press: *Figure 2.3:* Eisner, T. *et al.* (1969) *Science*, **166**, pp. 1172–1174, © American Association for the Advancement of Science; *Figure 2.4:* Lorenz, K. and Tinbergen, N. (1938) *Zeitschrift für Tierpsychologie*, **2**, pp. 1–29, Verlag Paul Parey, Berlin; *Figures 2.5a, 2.5b:* Kandel, E. R. (1976) *Cellular Basis of Behaviour*, copyright © 1976 by W. H. Freeman and Co., reprinted with permission; *Figures 2.5c, 2.11:* Manning, A. (1979) *An Introduction to Animal Behaviour*, 3rd edn, Edward Arnold Publishers; *Figure 2.6:* Halliday, T. R. (1974) *Journal of Herpetology*, **8**, pp. 277–292, Society for the Study of Amphibians and Reptiles; *Figure 2.8:* Crook, J. H. (1964) *Proceedings of the Zoological Society*, **142**, pp. 217–255, Zoological Society of London; *Figure 2.13:* Morris, D. (1970) *Patterns of Reproductive Behaviour*, Jonathan Cape; *Figure 2.15:* Farkas, S. R. and Shorey, H. H. (1972) *Science*, **178**, pp. 67–68, copyright © American Association for the Advancement of Science; *Figure 2.16:* Ingle, D. (1976) *The Amphibian Visual System*, Academic Press Inc; *Figures 2.17, 2.22, 2.23, 2.24, 4.1:* McFarland, D. (1981) *The Oxford Companion to Animal Behaviour*, Oxford University Press; *Figure 2.19:* Huntingford, F. A. (1984) *The Study of Animal Behaviour*, Chapman and Hall; *Figures 2.20, 2.21:* Dawkins, M. (1971) *Animal Behaviour*, **19**, pp. 566–574, Baillière Tindall; *Figure 2.25:* Slater, P. J. B. (1978) *Sex Hormones and Behaviour*, Edward Arnold Publishers; *Figure 3.5:* Gall, J. G., Cohen, E. H. and Polan, M. H. (1971) *Chromosoma*, **33**, pp. 319–344, Springer-Verlag; *Figure 3.9:* adapted from Grier, J. W. (1984) *Biology of Animal Behavior*, Mosby Year Book, Inc; *Figures 3.12, 3.14, 5.3:* Barnett, S. A. (1981) *Modern Ethology*, Oxford University Press; *Figure 3.15:* Strickberger, M. W. (1968) *Genetics*, Macmillan, New York, © Monroe W. Strickberger; *Figures 4.3, 4.10:* courtesy of Dr Marion Hall; *Figures 4.4, 4.9:* Clutton-Brock, T. H., Guinness, F. E. and Albon, S. D. (1982) *Red Deer: Behaviour and Ecology of Two Sexes*, reprinted by permission of University of Chicago Press; *Figure 4.8:* courtesy of Prof Paul W. Sherman/Cornell University; *Figure 4.12:* Eric and David Hosking Picture Collection; *Figure 4.13:* Niko Tinbergen/Dutch State Archives; *Figure 5.1:* Bonner, J. T. (1958) *The Evolution of Development*, Cambridge University Press; *Figure 5.6:* Hess, E. M. (1959) *Science*, **130**, pp. 133–141, copyright © 1959, by American Association for the Advancement of Science; *Figure 5.7:* Halliday, T. R. and Slater, P. J. B. (eds) (1983) *Animal Behaviour, 3: Genes, Development and Learning*, Blackwell Scientific Publications Ltd; *Figure 5.9:* Marler, P. (1981) *Trends In Neurosciences*, **4**, pp. 88–94, Elsevier Science Publishers; *Figure 5.10:*

Kroodsma, D. E. (1978) in G. Burghardt and M. Bekoff (eds) *The Development of Behavior: Comparative and Evolutionary Aspects*, Garland Publishing Inc, New York; *Figures 6.3, 6.4, 6.5, 6.6:* Tagawa, B. in E. R. Kandel (1970) *Scientific American*, **223**(1), pp. 57–70, reprinted by permission; *Figure 6.7:* Miller, R. R. and Marlin, N. A. (1981) *Journal of Experimental Psychology: Animal Behaviour Processes*, **7**, pp. 313–333, copyright © by the American Psychological Association; *Figure 6.9:* Wagner, A. R. *et al.* (1968) *Journal of Experimental Psychology*, **76**, pp. 171–180, copyright © 1968 by the American Psychological Association, adapted by permission; *Figure 6.12:* Moore, B. R. (1973) in R. A. Hinde and J. Stevenson-Hinde (eds) *Constraints on Learning: Limitations and Predispositions*, Academic Press Inc; *Figure 7.4:* Mower, G. D. (1976) *Journal of Comparative and Physiological Psychology*, **90**, pp. 1152–1155, and Cabanac, M. (1979) *Journal de Physiologie, Paris*, **75**, pp. 115–178; *Figure 7.6:* Mook, D. (1987) *Motivation: The Organization of Action*, by permission of W. W. Norton and Co., Inc., copyright © 1987 by W. W. Norton and Company, Inc; *Figure 7.7:* Mook, D. G. (1963) *Journal of Comparative and Physiological Psychology*, **56**, pp. 645–659, copyright © by the American Psychological Association; *Figure 7.9:* Oatley, K. and Toates, F. M. (1969) *Psychonomic Science*, **16**, pp. 225–226, reprinted by permission of the Psychonomic Society; *Figure 7.10:* Michael, R. P. and Zumpe, D. (1981) in D. Gilmore and B. Cook (eds) *Environmental Factors in Mammalian Reproduction*, Macmillan; *Figure 7.11:* adapted from Toates, F. (1986) *Motivational Systems*, Cambridge University Press; *Figure 7.12:* Hogan, J. A. *et al.* (1970) *Journal of Comparative and Physiological Psychology*, **70**, pp. 351–357, Springer-Verlag Inc; *Figures 8.1, 8.4:* Bolles, R. C. (1979) *Learning Theory*, Holt Rinehart and Winston, reprinted by permission of the Regents of the University of California; *Figure 8.2:* Toates, F. (1986) *Motivational Systems*, Cambridge University Press; *Figure 8.3a:* O'Keefe, J. and Nadel, L. (1978) *The Hippocampus as a Cognitive Map*, Oxford University Press; *Figure 8.5:* Archer, J. and Birke, L. (1983) *Exploration in Animals and Humans*, Van Nostrand Reinhold; *Figures 8.10, 8.11, 8.12:* Gallup, G. G. (1977) *American Psychologist*, **32**, pp. 329–338, copyright © American Association for the Advancement of Science; *Figure 8.13:* Gallup, G. G., McClure, M. K., Hill, S. D. and Bundy, R. A. (1971) *The Psychological Record*, **21**, pp. 69–74, copyright © 1971 by the Psychological Record, reprinted by permission; *Figure 9.1:* Campbell, B. and Lack, E. (1985) *A Dictionary of Birds*, British Ornithologists' Union; *Figure 9.2:* Slater, P. J. B. (1978) *Sex Hormones and Behaviour*, Edward Arnold Publishers; *Figure 9.3:* Eberhard, W. G. (1985) reprinted by permission of the publishers, from *Sexual Selection and Animal Genitalia*, Harvard University Press, Cambridge, Mass., copyright © 1985 by the President and Fellows of Harvard College; *Figure 9.4:* Mock, D. W. and Fujioka, M. (1990) *Trends in Ecology and Evolution*, **5**, pp. 39–43, Elsevier Science Publishers; *Figure 9.5:* Baerends, G. P., Brouwer, R. and Waterbolk, H. T.(1955) *Behaviour*, **8**, pp. 249–334, E. J. Brill; *Figure 9.6:* courtesy of Dr Tim Halliday; *Figure 9.8:* Andersson, M. (1982), reprinted by permission from *Nature*, **299**, pp. 818–820, copyright © 1982 Macmillan Magazines Ltd; *Figure 9.9:* Hinde, R. A. and Simpson, M. J. A. (1975) in *Ciba Foundation Symposium 33, Parent–Infant Interactions*, Elsevier Science Publishers; *Figure 10.1:* Andersson, M. and Wiklund, C. G. (1978) *Animal Behaviour*, **26**, pp. 1207–1212, Baillière Tindall; *Figure 10.2:* Bertram, B. C. R. (1980) *Animal Behaviour*, **28**, pp. 278–286, Baillière Tindall; *Figure 10.3:* Hamilton, W. D. (1971) *Journal of Theoretical Biology*, **31**, pp. 295–311, Academic Press; *Figure 10.4:* Major, P. F. (1978) *Animal Behaviour*, **26**, pp. 760–

777, Baillière Tindall; *Figure 10.5:* Elgar, M. (1986) *Animal Behaviour,* **34**, pp. 169–174, Baillière Tindall; *Figure 10.6:* Pulliam, H. R. (1976) in P. H. Klopfer and P. P. G. Bateson (eds) *Perspectives in Ethology,* Vol. 2, Plenum Publishing Corporation, New York; *Figure 10.9:* Fitzpatrick, J. W. and Woolfenden, G. E. (1988) in Clutton-Brock, T. H. (ed.) *Reproductive Success,* University of Chicago Press, copyright © by the University of Chicago; *Figure 10.10:* Zoo Operations Ltd; *Figure 10.11:* Wilkinson, G. S. (1984), reprinted by permission from *Nature,* **308**, pp. 181–184, copyright © 1984 Macmillan Magazines Ltd.

INDEX

neutral, 58
mutualism, 259–61

naked mole rat, eusociality, 258–9
natural selection, 77–91, **79**
   directional, 88
   disruptive, 88
   stabilizing, 85, *87*
   see also kin selection; group selection
nature–nurture debate, 13
negative feedback, **170**–71, 174, 186
negative reinforcement, **151**–2, 157
Nemeritis canescens see under wasps
nesting behaviour
   kittiwake, 97–8, 99
   lovebird, 64
   swallow-tailed gull, 98–9
   weaverbird, 27–8
neutral mutation, **58**
neutral stimulus (NS), **146**
newt *see* smooth newt
non-associative learning, **139**, 140–5
non-homeostatic systems, 175–6
Nucifraga caryocatactes (nutcracker) memory, 200
Nucifraga columbiana (Clark's nutcracker), food-storing
   behaviour, 135
nutcracker
   memory, 200
   see also Clark's nutcracker

observation, 9
octopus (*Octopus vulgaris*), sensitization in, 143
oddity problems, 156
odour plume, 36, *37, 38*
oestrogen, **45**, 175, 217, *218*
olfactory imprinting, 127
olive baboon, self-recognition not shown by, 209
operant, **150**
operant conditioning, **150**–3
orange-rumped honeyguide, resource-based polygyny,
   224, 230
orientation, 31, 36–7
Oryctolagus cuniculus (rabbit)
   conditioning of eye-blink response, 149
   reproductive potential, 78
ostrich, group vigilance, 244

Pan troglodytes (chimpanzee)
   behaviour showing possible insight, 154
   secondary reinforcement of behaviour, 153

self-recognition, 208–9
social behaviour, 205–6
Papio cynocephalus (olive baboon), self-recognition not
   shown by, 209
Paralastor see under wasp
parental care, 220, 221–3, 235–7
parental investment, 220
Parrish, J., 245
parthenogenesis, 219
Parus ater (coal tit), memory in, 159
Parus caeruleus (blue tit)
   learning behaviour, 137–8
   male-removal experiments, *229*
   memory in, 159
   transmission of information, 262
Parus major (great tit)
   function of song, 11–12
   male-removal experiments, *229*
   memory in, 159
   timing of breeding, 216
Parus palustris (marsh tit), memory in, 159
Passer domesticus (house sparrow)
   badges of status, 255
   chirrup calls, 247–8
Paul, R., 65
Pavlov, I., 146–8, 157
Pavo cristatus (peacock), mate choice, 232–3
peach-faced lovebird, nest-building behaviour, 64
peacock, mate choice, 232–3
pecking order, 254
peripheral filtering, **39**
Petrie, M., 232
phenocopies, **72**
phenotype, **55**
phenylalanine hydroxylase, 62–3, 114
phenylketonuria (PKU), **62**–3, 70–1
pheromone, **36**
Philanthus triangulum (digger wasp), orientation, 31
Philomachus pugnax (ruff), variation in behaviour, 76,
   102–3
Photinus (firefly), flash patterns, 31
Photurus (firefly), flash patterns 31
phylogeny of behaviour, **6**
pied flycatcher, male-removal experiments, *229*
pig, persistent rooting behaviour, 156
pigeon
   activity switching, 185–6
   flight learning, 111, 119
   male-removal experiments, *229*
   see also wood pigeon

reproductive potential, 77–8
of males and females, 221
reproductive success *see* lifetime reproductive success
reproductive synchrony, 222–3
response, **17**
conditional, 146
unconditional, 146
rhesus monkey
effects of social deprivation, 117
learning behaviour, 155–6
mother–infant relationships, *237*
self-recognition not shown by, 209
sexual motivation 175–6
social behaviour, 204–5
Rissa tridactyla (kittiwake)
breeding success, 235
nesting behaviour, 97–9
pair-bonds, 224
robin, attack response, 19
Rohwer, S., 227
Rothenbuhler, W., 59–61
ruff, variation in behaviour, 76, 102–3

Saguinus labiatus (tamarin), vigilance, 244
Saimiri sciureus (squirrel monkey), social relationships, 244
salamander *see* tiger salamander; two-lined salamander
salmon (*Salmo salar*), olfactory imprinting, 127
satellite male (ruff), 76, 102–3
Savage-Rumbaugh, S., 206
sea slug, escape response, 23, *24–5*
seabass *see* black seabass
seahorse, mating and parental care, 221
search image hypothesis, 40–1
seaside sparrow, male-removal experiments, 228, *229*
second-order conditioning, 147
secondary reinforcement, 153
sedge warbler, song, 11
selection *see* natural selection; kin selection; group selection
selection coefficient, **80–2**
selection pressure, **80**, 84
selective attention, 39–40
selective breeding, for behavioural characters, 66–7
self-recognition, 207–9
selfish herd concept, 245
sense organs, and behaviour, 18–21
sense (sensory modality), **20–1**
sensitive period, **13**, 124
song learning, 13, 130

sensitization, **143–4**
sensory modality *see* sense
sequences of behaviour, 25–7
Serinus canaria (canary), song, 131
Sevenster-Bol, A. C. A., 34
sex
concept of, 218–19
example of polymorphism, 76
sex chromosomes, 56, 72
sex hormones, 43, 45, 175–6
sex-role reversal, 220–1
sexual behaviour, 115–16, 126, 167, 215
mechanisms, 216–18
sexual imprinting, **125**–6, 236
sexual motivation, 164, *165*, 167, 175–6
Coolidge effect, 178
sexual reproduction, 219
Seyfarth, R., 206–7
shaft-tailed widowbird, mate choice, 235
shaping (in conditioning experiments), **150**
sheepdog, training, 262
Sherman, P. W., 90–1, 258
short-day species, **216**–17
Siamese fighting fish, response to rewards, 178–9
sign stimulus, **28**
silk moth, pheromone, 38, 42
Silver, R., 217
Skinner, B. F., 150, 152–3
Skinner box, 150–3, 157, 164, 175, 183–4, 186, 201–2
smooth newt, courtship behaviour, 9–11, 26–7
snails
feeding, 21, 24
habituation in behaviour, 140
orientation, 36
snake, sensitive to infrared radiation, 38
snow bunting, male-removal experiments, *229*
social behaviour, 241–63
social conflict, 242, 254–5
social deprivation, 117
social relationships, 241–2
somatic effort, **220**
sonagram, **64**, *65*, 129, *130, 131*
song sparrow, song learning, 128, 130
sparrow *see* hedge sparrow; house sparrow; seaside sparrow; song sparrow; white-crowned sparrow; white-throated sparrow
species, **4**
species-specific behaviour, 118–21, 167
species-specific characteristics, **13**, 21, 41–2
Spencer-Booth, Y., 236